Practical Solid-State Circuit Design

by

Jerome E. Oleksy

Howard W. Sams & Co., Inc.
4300 WEST 62ND ST. INDIANAPOLIS, INDIANA 46268 USA

Copyright © 1974 by Howard W. Sams & Co., Inc.,
Indianapolis, Indiana 46268

FIRST EDITION
THIRD PRINTING—1977

All rights reserved. Reproduction or use, without express permission, of editorial or pictorial content, in any manner, is prohibited. No patent liability is assumed with respect to the use of the information contained herein. While every precaution has been taken in the preparation of this book, the publisher assumes no responsibility for errors or omissions. Neither is any liability assumed for damages resulting from the use of the information contained herein.

International Standard Book Number: 0-672-21018-5
Library of Congress Catalog Card Number: 73-90284

Printed in the United States of America.

Preface

Few things are more rewarding in electronics than successfully designing circuits to meet your own requirements. Essentially, there are two ways of designing electronic circuits: the analytical (mathematical) method, and the experimental method. In this book, the latter approach will be emphasized.

It has been said that experimentation is the true test of knowledge. For the electronics man, it is also the fun part of knowledge. I hope you will proceed through the book, trying out the techniques outlined, and see for yourself that learning to design can be a lot of fun.

In each of the chapters of this book, we will first look at some simplified explanations of how the circuits and components work. Next, we will outline step-by-step design procedures, with sufficient explanation on how to choose appropriate components to build the circuits. All you need at the start is some knowledge of Ohm's law for ac and dc circuits and an understanding of how to use simple test equipment, such as a vom, an oscilloscope, and an ac signal generator.

At various points throughout the text you will find short examples and quizzes. Some of the quizzes involve calculations from equations given in the text; others call for qualitative answers, such as "If R_2 in the circuit is decreased, the voltage at point A will—[increase, decrease, remain the same]." By performing these "mental experiments" and comparing your answers with those given in the back of the book, you will be able to gauge your understanding of the subject.

By the time you finish this book, you will be able to design many circuits "from scratch," using easily obtainable components. You will also have gained some valuable insight into how to modify existing circuits experimentally to meet your own needs.

I wish to express my sincere thanks to George B. Rutowski, P.E., for innumerable helpful comments and criticisms. Also a special thanks to my wife, Betty, for typing this manuscript and for her patience.

JEROME E. OLEKSY

Contents

CHAPTER 1

POWER SUPPLIES 7
    Solid-State Diodes—Half-Wave Rectifier—Full-Wave Rectifier—
Full-Wave Bridge—Summary of Design Procedure

CHAPTER 2

TRANSISTOR AMPLIFIER DESIGN (SINGLE-STAGE) 29
    Bipolar Transistors—Biasing—Amplification of AC Signals—
Voltage Divider Bias—Temperature Effects—Summary

CHAPTER 3

TRANSISTOR AMPLIFIER DESIGN (CASCADED STAGES) . . 47
    Input Impedance—Emitter Followers—Feedback in Amplifiers—
Frequency Response of RC-Coupled Stages—Gain Controls—
Miscellaneous Design Pointers

CHAPTER 4

FIELD EFFECT TRANSISTORS 75
    JFET Characteristics—The JFET Amplifier—MOSFETs—FET
Applications—Summary—Glossary of FET Terms

CHAPTER 5

DIFFERENTIAL AND OPERATIONAL AMPLIFIERS 93
    Differential Amplifiers—Operational Amplifiers—Ideal Op Amp—
Inverting Amplifier—Noninverting Amplifier—Voltage Follower
—Summer—Integrator—Differential-Input Amplifier—Summary

CHAPTER 6

USING INTEGRATED CIRCUIT OP AMPS 113

Output Offset Caused by Input Offset Voltage—Output Offset Caused by Input Bias Current—Frequency Response—Pointers on Working With IC Op Amps—Summary—Glossary of Op Amp Terms

CHAPTER 7

OSCILLATORS AND WAVEFORM GENERATORS 127

Phase-Shift Oscillator—Twin T Oscillator—Square-Wave Generator—Converting Sine Waves to Square Waves—Converting Square Waves to Triangular Waves—Summary

CHAPTER 8

AUDIO POWER AMPLIFIERS 143

Complementary-Symmetry Amplifiers—Heat Sinks—Pointers on Constructing Power Amplifiers—Summary

CHAPTER 9

REGULATED POWER SUPPLIES 157

Zener Diode Regulator—Higher Current Regulators—Variable Voltage Regulators—Current Limiters—Summary

APPENDIX

QUIZ ANSWERS 175

INDEX 185

1

Power Supplies

Nearly every electronic circuit requires some sort of power source or power supply. An amplifier, whether for hi-fi or digital applications, is basically an energy converter, used in most cases to convert dc power into ac signal power. The most common form of power source is the dc power supply, which can be either a battery or a rectifier circuit which changes ac from the power line into suitable dc. Batteries are well suited for portable equipment, but they eventually run down. For applications requiring considerable current, or for applications where batteries would be inconvenient, the rectifier-type power supply is the most practical solution. In this chapter we will examine a few of the more common rectifier power supplies used in solid-state electronic equipment. These have a relatively low power consumption, with voltages on the order of 100 volts or less, and currents of up to 1 or 2 amps.

The difference between ac and dc is that alternating current periodically reverses direction, whereas direct current flows only in one direction. In order to change ac into dc, we need a device that conducts current in only one direction. A *diode* is such a device.

SOLID-STATE DIODES

The solid-state diode is formed by joining two different types of semiconductor material together. Semiconductors are materials which are not good conductors, as is copper, but neither are they good insulators, as is glass. The two types of semiconductor material, called n-type and p-type, are formed by putting controlled amounts of impurities such as arsenic, gallium, indium, etc., into pure crystals of either germanium (Ge) or silicon (Si). If some arsenic atoms are put

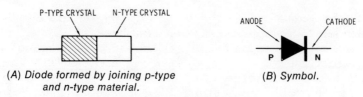

(A) Diode formed by joining p-type and n-type material.

(B) Symbol.

Fig. 1-1. Solid-state diode.

into molten silicon during the manufacture of silicon crystal, an n-type crystal is formed. This means that the crystal has negative charge carriers (electrons) which are free to move about the crystal. On the other hand, if indium atoms are put into the molten silicon, p-type crystal is formed, which has positive charge carriers (holes) that are free to move about.

There are many good solid-state physics texts which explain how and why the diode operates and how the holes and electrons move about. However, from the practical viewpoint, it will generally be sufficient to know that the solid-state diode is made up of two different types of semiconductor material—n-type and p-type—joined together as shown in Fig. 1-1A. The schematic symbol for the solid-state diode is shown in Fig. 1-1B. The p-type material of the diode is called the anode and the n-type material is called the cathode. The joining together of n- and p-type materials produces a *junction diode*.

As stated previously, a diode conducts current only in one direction, when the anode is made positive with respect to the cathode as shown in Fig. 1-2A. The arrow on the diode symbol points in the direction of conventional current flow. When the diode is conducting, it is said to be *forward-biased*. As shown in Fig. 1-2B, the diode does not conduct when the anode is negative with respect to the cathode. The diode is then said to be *reverse-biased,* or *back-biased*.

Diodes come in various cases, or packages, a few of which are shown in Fig. 1-3. Note that many types have the cathode end marked with a stripe or band. The larger cases are for diodes capable of carrying higher currents. Diode identification numbers usually begin with 1N, such as 1N914, 1N4001, etc. Although diodes are made

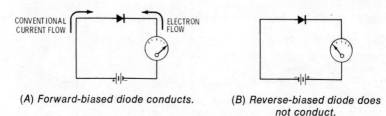

(A) Forward-biased diode conducts.

(B) Reverse-biased diode does not conduct.

Fig. 1-2. Diode operation.

from either germanium (Ge) or silicon (Si), we will limit our discussion here to *rectifier* diodes, which are usually silicon.

A simple ohmmeter can be used to demonstrate "diode action"—that is, conducting when forward-biased and not conducting when reverse-biased. If an ohmmeter is connected to a diode as in Fig. 1-4A, the internal battery of the ohmmeter forward-biases the diode,

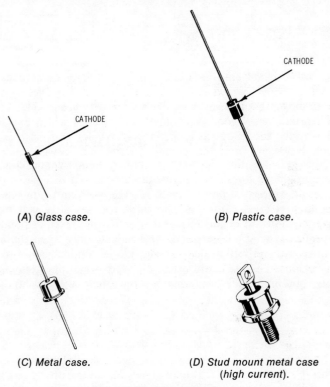

(A) Glass case.

(B) Plastic case.

(C) Metal case.

(D) Stud mount metal case (*high current*).

Fig. 1-3. Diode packages.

causing current to flow. You then read a low value of resistance. By reversing the leads of the ohmmeter (reverse-biasing the diode), practically no current will flow, and the result is a very high resistance, as shown in Fig. 1-4B. The actual values of resistance measured mean very little; the important thing is that a low resistance is measured in one direction, and a high resistance in the other. This test also gives you a simple way of determining which end of the diode is the cathode and which is the anode, in the event they are not marked. Of course, you must first be sure which lead of the ohmmeter is connected to the negative terminal of the internal battery; most often the terminal marked "common" is negative.

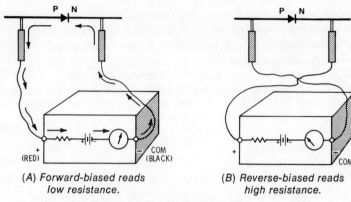

(A) Forward-biased reads low resistance.

(B) Reverse-biased reads high resistance.

Fig. 1-4. Demonstrating diode action with an ohmmeter.

HALF-WAVE RECTIFIER

Now let's use the diode as a rectifier. Suppose you connect the diode in a circuit with an ac generator, as shown in Fig. 1-5A. Measuring the voltage at point A with respect to ground, you read the ac generator voltage. (Say the generator produces a 12-volt peak sine wave, as shown in Fig. 1-5B). But what if you looked at point B with respect to ground? Since the diode conducts only when the anode is positive with respect to the cathode, it will conduct only on *positive* alternations. That is, it will act as a closed switch on positive alternations, but on negative alternations the diode will act as an open

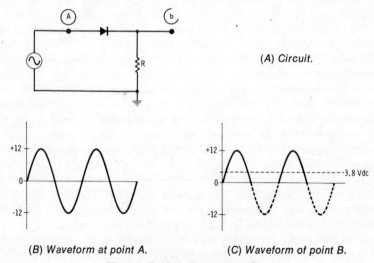

(A) Circuit.

(B) Waveform at point A.

(C) Waveform of point B.

Fig. 1-5. Basic half-wave rectifier.

switch. No current can flow through R on the negative alternations, so the waveform across R will look like Fig. 1-5C. When used in this manner, the diode is called a *half-wave rectifier,* because it conducts on half of the input wave.

The waveform across R has a dc equivalent. That is, if you connect a dc voltmeter across R you will read some dc voltage, but if you connect it to point A from ground it will read zero volts dc; so the diode changes ac into dc. The value of this equivalent dc voltage (V_o) is about 0.318 times the peak voltage (V_p) of the input sine wave. In this case, V_p is 12 volts, making V_o approximately 3.8 Vdc.

Although we have said that the diode looks like a closed switch when forward-biased, this is true only for a theoretically ideal diode. Practical diodes actually have a slight voltage drop across them when conducting. This voltage, called V_f (forward voltage drop), is about 0.3 volts for Ge (germanium) diodes and 0.7 volts for Si (silicon) diodes. In Fig. 1-5C, for example, the positive peaks appearing across R would actually reach a value of 11.3 volts if you were using a Si diode. This small voltage drop across the conducting diode can often be ignored, but if you are working with fairly low voltages (say less than 10 volts peak), you may need to take it into account.

Although we have changed ac into dc, we still have pulsating, or varying, dc. Most electronic applications need smooth, or constant-value, dc. The simplest way of smoothing out the pulses is by connecting a large *filter* capacitor across R as shown in Fig. 1-6.

Here's how the filter circuit works. Assume that the capacitor is initially uncharged. The generator is then turned on and current flows during the positive alternation, charging up the capacitor and also developing a voltage across resistor R. This occurs from times t_0 to t_1 in Fig. 1-6C. After time t_1, the voltage at A starts to decrease, as shown by the dotted sine wave, but the capacitor is still charged. When the voltage at A drops to a value less positive than the voltage across capacitor C, the diode becomes reverse-biased and nonconducting. The capacitor then sees only R across it and starts discharging through R as shown in Fig. 1-6B. As long as the capacitor is discharging through R, some voltage appears across R between times t_1 and t_2, as shown in Fig. 1-6C.

Notice that at time t_2 the input voltage starts to exceed the voltage across the capacitor. When this happens, the diode once again becomes forward-biased and causes the capacitor to recharge to the input voltage as shown between t_2 and t_3. After t_3 the diode again becomes reverse-biased, the capacitor discharges once more, and the process repeats itself over and over.

Thus the purpose of connecting the filter capacitor across R is to hold the voltage across R more nearly constant, and at a higher average value. As shown in Fig. 1-6C, the average (dc) voltage (V_{av}) now

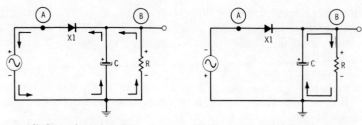

(A) Capacitor charging. (B) Capacitor discharging through R.

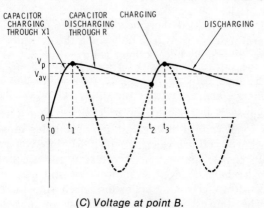

(C) Voltage at point B.

Fig. 1-6. Filtering rectified ac.

appearing across R is approximately midway between the peak value and the minimum capacitor voltage.

There is still some ripple across R; that is, the dc voltage still fluctuates somewhat. A larger capacitor holds the charge longer, and therefore does not discharge as much between charging peaks (Fig. 1-7). Your challenge as a designer is to choose a capacitor large enough to minimize the ripple to a reasonable value, but not too much larger than necessary so as to save on cost and physical size.

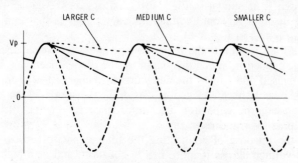

Fig. 1-7. Effect of capacitance values on filtering.

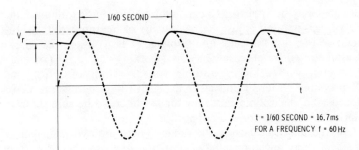

Fig. 1-8. Determining the pulse interval.

Obviously the smaller R is, the faster the capacitor will discharge, because the discharge current through R will be greater. Also, the longer the time between recharging pulses, the more the capacitor will discharge, and the greater the ripple. These relationships can be used to calculate the desired value of capacitance. We must know the discharge current and the time between recharging pulses. As can be seen in Fig. 1-8, the time between recharging pulses is about 1/60 of a second, or 16.7 ms for a 60-Hz input frequency. We can find the value for C by using the formula,

$$C = \frac{I \times t}{V_r}$$

where,
C is the capacitance in farads,
I is the dc current drawn by R in amps,
t is 16.7 ms (the period of the wave),
V_r is the peak-to-peak ripple voltage in volts.

Exactly how much ripple voltage can be tolerated depends on the application. A single filter capacitor can reduce ripple to a few tenths of a volt, but if that is still too much you can use additional filtering or a regulator. This will be discussed in a later chapter.

EXAMPLE 1-1—For a half-wave rectifier like that of Fig. 1-6A, find the value of a filter capacitor, assuming that you want no more than 0.5 volts peak-to-peak ripple and the dc current drawn by R is 20 mA.

SOLUTION

$$C = \frac{I \times t}{V_r} = \frac{20 \times 10^{-3} \times 16.7 \times 10^{-3}}{0.5} = 670 \ \mu F$$

In practice you can use the closest standard value of capacitance.

When buying a capacitor, always specify the working voltage as well as the capacitance. The working voltage is simply the peak voltage that can safely be applied to a capacitor continuously without damage. In a power supply it is just the peak value of the input voltage. In the circuit of Fig. 1-6 the peak input voltage is 12 volts, so the capacitor should be rated for at least 12 Vdcw (working volts dc). You can use a capacitor with a higher working voltage rating, of course, but it will probably be larger and more expensive than necessary. You will often use an electrolytic capacitor for the filter, so be sure the correct polarity is observed for the connection.

Let's go now to selecting the diode. There are two main points to consider when selecting a rectifier diode: the amount of current it must carry, and the peak reverse voltage that will appear across it. Manufacturers list values of average forward current and prv (peak reverse voltage), also labeled piv (peak inverse voltage). For the diode in the previous example, the average rectified current is simply the current drawn by the load, or 20 mA. Any diode capable of withstanding at least 20 mA forward current could be used. You could certainly use a diode with a larger current rating, say 1 amp or so. In fact, since 1-amp diodes are used very often, they are manufactured in such large quantities that it might even be cheaper to buy a 1-amp diode rather than one with a lower current rating.

Next determine what maximum voltage will appear across the diode. From Fig. 1-8 you will see that, when the input voltage (dotted sine wave) is at a negative peak (-12 volts), the voltage across the capacitor is still near the positive peak of $+12$ volts. At this instant, the total voltage across the reverse-biased diode is the sum of the capacitor voltage and the input voltage, or 24 volts. That is, in a half-wave rectifier the peak reverse voltage appearing across the diode is *twice* the peak input voltage. For a power supply with a negative output, simply reverse the polarity of the diode and filter capacitor, as shown in Fig. 1-9B.

A transformer is ordinarily required to step up or step down the line voltage to the desired value for your supply. (Transformer con-

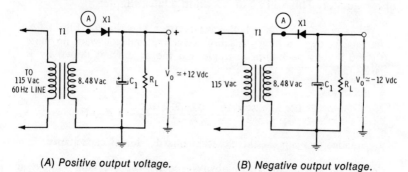

(A) *Positive output voltage.*   (B) *Negative output voltage.*

Fig. 1-9. Rectifier with transformer input.

nections are shown in Fig. 1-9.) Transformers are made for a wide variety of applications, but here we are interested mainly in rectifier power transformers. Many times, filament transformers are excellent for making low-voltage power supplies for solid-state work. Two major specifications to consider in selecting a transformer are the secondary voltage and secondary current ratings. The secondary voltage is given as an rms value, so depending on the peak secondary voltage you want, convert that value to rms and select a suitable transformer. For example, to replace the generator of Fig. 1-5 with a transformer you would need one with a secondary voltage of

$$V_{rms} = 0.707 \, V_p$$
$$= 0.707 \times 12 = 8.48 \text{ volts.}$$

It is unlikely that a transformer with this exact value of secondary voltage will be available; choose one with the closest standard value. One manufacturer, for example, lists transformers with secondary voltages of 7.5 volts and 10 volts. You could use either of these if the output voltage value is not critical. You may also be concerned with cost, and sticking to commonly used voltages, such as 6.3 volts or 12.6 volts (used for tube filaments), is usually the most economical. In a later chapter we will study variable-voltage power supplies that you can use whenever the output voltage is critical.

Some manufacturers also list the maximum secondary dc current, and as long as the transformer current rating is slightly higher than the anticipated load current, you're all set. If the secondary rms current is listed instead, multiply the dc current by a form factor (Table 1-1) to get the rms current.

A few other components you may want to include in your final supply are a line plug, switch, fuse, and power-on indicator lamp, as shown in Fig. 1-10. To choose the correct size for the fuse, you should find the average primary current. If the transformer is used to step the voltage down, the primary current will be less than the secondary current. The standard transformer equations apply:

$$\frac{V_p}{V_s} \cong \frac{N_p}{N_s} \cong \frac{I_s}{I_p}$$

where,
  V_p is the primary voltage,
  V_s is the secondary voltage,
  N_p is the number of turns in the primary coil,
  N_s is the number of turns in the secondary coil,
  I_p is the primary current,
  I_s is the secondary current.

Often a fuse is inserted to protect the transformer in the event that the secondary becomes shorted. (It can be slightly higher than the

Table 1-1. Power Supply Design Requirements

Component	Basic Requirement	How to Determine Requirement			Other Useful Information in Catalog
		Half Wave (Fig. 1-9)	Full Wave (Fig. 1-12)	Bridge (Fig. 1-14)	
Transformer	(1) $V_s =$	$0.707 V_o$	$1.41 V_o$	$0.707 V_o$	Cost
	(2) $I_s =$	$\geq 2.3 I_o$	$\geq 1.2 I_o$	$\geq 1.8 I_o$	Size
					Insulation
Diodes	(3) $I_f =$	$\geq I_o$	$\geq 0.5 I_o$	$\geq 0.5 I_o$	Cost
	(4) prv $=$	$\geq 2 V_o$	$\geq 2 V_o$	$\geq V_o$	Mounting
					Operating Temp.
Filter Capacitor	(5) C $=$	$\geq \dfrac{It}{V_r}$ (t = 16.7 ms)	$\geq \dfrac{It}{V_r}$ (t = 8.35 ms)		Cost
	(6) V dcw $=$	$\geq V_o$	$\geq V_o$		Mounting
					Operating Temp.

$V_o =$ dc output voltage in volts,
$I_o =$ dc output current in amps,
$V_r =$ ripple voltage in volts (peak-to-peak),
prv $=$ peak reverse voltage in volts,
V dcw $=$ working volts dc in volts,
$V_s =$ transformer secondary voltage in volts rms,
$I_s =$ transformer secondary current in amps.

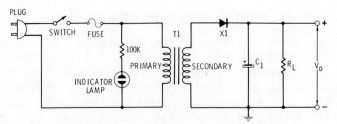

Fig. 1-10. Complete half-wave rectifier supply.

transformer current capacity.) A fuse acts too slowly to protect solid-state components, so its function here is solely to protect the transformer. In a later chapter, we will study current-limiting circuits which can be used in power supplies to protect the solid-state devices.

The plug and switch must be rated for 115-volt line operation and must be able to handle the primary current. Indicator lamps of various types can be found in catalogs. One popular type is a neon lamp indicator, which requires a resistor in series (about 100 K ohms) to limit the lamp current. These resistors are often included in the base of the lamp when you buy it.

Quiz

1. Suppose you have a 12.6-volt rms filament transformer, rated at 1.5 amps secondary current, with a primary designed for 115 Vac, 60 Hz. You want to build a half-wave rectifier as in Fig. 1-9. Load resistor R_L will draw 50 mA, and you want no more than 1 volt peak-to-peak ripple across it.

    (A) What will be the peak input voltage?
    (B) What capacitance value should you use?
    (C) What should be the minimum voltage rating for the capacitor?
    (D) What should be the minimum current rating for X1?
    (E) What should be the prv rating for X1?

2. Suppose you build the power supply in Problem 1. Predict the results if the changes stated were made in the circuit by answering I (increase), D (decrease), or S (remain the same) to the following.

    (A) If the capacitance is reduced by 50%, the ripple voltage (V_r) will _____.
    (B) If the capacitance is reduced by 50%, V_o will _____.
    (C) If R_L is decreased by 50%, the load current will _____.
    (D) If R_L is decreased by 50%, V_r will _____.
    (E) If R_L is decreased by 50%, V_o will _____.
    (F) If the capacitor becomes leaky, V_r will _____.

FULL-WAVE RECTIFIER

As we saw in the previous section, the filter capacitor in a power supply must maintain a voltage across the load between recharging pulses. If the recharging pulses occur more frequently, a smaller capacitor can be used for the same load current and not have excessive ripple. One way to make the recharging pulses occur more frequently is to use *both* alternations of the input cycle to recharge the capacitor. Such a power supply, called a *full-wave rectifier* supply, is shown in Fig. 1-11. Fig. 1-11A shows the alternation when the secondary voltage at point A is positive with respect to ground (center tap). During this alternation, diode X1 is forward-biased and current flows as shown by the arrows. During the other input alternation (Fig. 1-11B), the bottom of the transformer (point B) is positive with respect to ground, and diode X2 conducts. Notice that on both alternations, current flows through R_L in the same direction. The resultant voltage across R_L is shown in Fig. 1-11C. The complete circuit, with capacitor filtration, is usually drawn as in Fig. 1-12A. The voltage across load resistor R_L will look like Fig. 1-12B. Note that the capacitor still maintains the voltage across R_L between recharging pulses, but here the pulses occur twice as often as in the half-wave circuit. The value for C can be determined as before, but the time interval is now 8.35 ms instead of 16.7 ms. As before, the prv rating for the diodes is twice the peak input voltage, but the average current through the

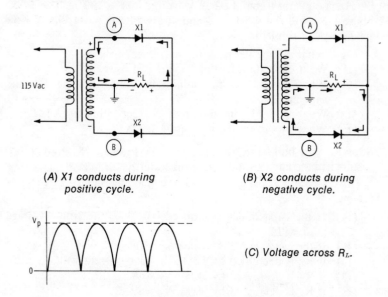

(A) *X1 conducts during positive cycle.*

(B) *X2 conducts during negative cycle.*

(C) *Voltage across R_L.*

Fig. 1-11. Full-wave rectifier.

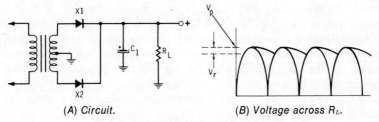

(A) Circuit.   (B) Voltage across R_L.

Fig. 1-12. Complete full-wave rectifier.

diodes is only half as great as in the half-wave circuit. This is because each diode need only replace half the charge drained from C for the entire cycle each time it conducts. As with the half-wave rectifier circuit, you can get an output voltage which is negative with respect to ground by reversing the polarities of the diodes and the capacitor.

FULL-WAVE BRIDGE

The full-wave rectifier circuit of Fig. 1-12A has a drawback: it requires a center-tapped (ct) transformer. Only half of the secondary is being used at a time, so it must have twice as many turns, and, consequently, greater bulk than is necessary. We can eliminate the center-tapped transformer by using a different type of circuit.

Probably the most commonly used type of power supply today is the *full-wave bridge* rectifier supply (Fig. 1-13A). Fig. 1-13B shows that diodes X2 and X3 are both forward-biased when the top of the secondary is positive with respect to the bottom. Electron flow is as

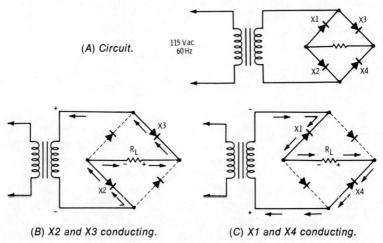

(A) Circuit.

(B) X2 and X3 conducting.   (C) X1 and X4 conducting.

Fig. 1-13. Full-wave bridge.

shown by the arrows. When the bottom of the secondary is positive with respect to the top (Fig. 1-13C), diodes X1 and X4 are forward-biased. Again current flows through R_L to the right. Hence both alternations produce the same polarity of voltage across R_L; this is full-wave rectification. The cost of the extra diodes is usually very low. Semiconductors have become very inexpensive in the last few years, and the cost savings on the smaller transformer usually offsets the added cost of the diodes.

Another interesting difference between a bridge-rectifier circuit and one using a center-tapped transformer is that the diodes for the bridge circuit need only have a prv rating equal to the peak of the input voltage, instead of twice the peak. In Fig. 1-13B, when the top of the transformer is positive, the voltage across diode X1 (non-conducting) is equal to the voltage across R_L plus the voltage across X3. Since the drop across X3 is very small, the peak voltage across X1 is essentially the peak voltage across R_L—that is, V_p. Similarly, the reverse voltage across X4 is equal to the voltage across R_L plus the voltage across X2, or approximately V_p. By similar reasoning, the prv across X2 and X3 on the other half-cycle is just V_p. In other words, you can use diodes with a smaller prv rating in the bridge-rectifier circuit than you can in the full-wave circuit, and lower prv-rated diodes are usually less expensive. Fig. 1-14 shows the full-wave bridge circuit drawn in the more conventional manner, complete with filter capacitor C_1 connected across the load.

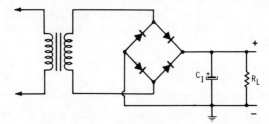

Fig. 1-14. Full-wave bridge with filter capacitor.

As in the other supplies, the dc output voltage (V_o) from this bridge-type supply will be approximately equal to the peak of the input voltage, minus the drop across the conducting diodes. In this case, there are always two conducting diodes in series, so the output will be about 1.4 volts less than V_p. Technically, the effective dc output is still slightly less than this, due to the small ripple voltage (V_r). It should be clear from Figs. 1-8 and 1-12B that the filtered dc output (V_o) can never be quite as great as V_p, even accounting for diode voltage drops. To accomplish this would require that V_r be exactly zero. However, V_r is generally such a small fraction (a few percent) of

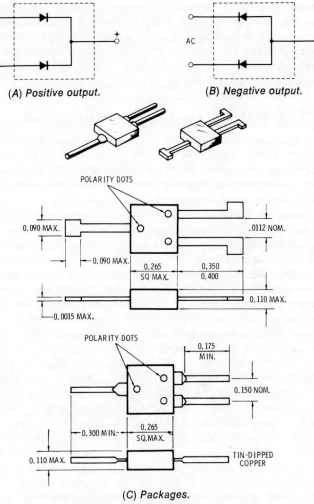

(A) Positive output. (B) Negative output.

(C) Packages.

Fig. 1-15. Full-wave rectifier assemblies.

V_o that it can be neglected in calculations. For most purposes, the dc output is simply V_p minus the total voltage drops of the diodes; for output voltages much larger than 10 V, a further approximation is to ignore diode effects and say that $V_o \cong V_p$.

Since full-wave power supplies are so common, manufacturers offer special rectifier assemblies with either two or four diodes internally connected and only three or four external connections (Fig. 1-15). Fig. 1-15A shows two diodes in a package that could be used to build the supply of Fig. 1-12. Fig. 1-15B shows two diodes in a similar connection, but wired for a negative output voltage with respect to ground. Fig. 1-16 shows a full-wave bridge rectifier assembly. Either positive or negative output voltages can be obtained from this unit merely by grounding the minus or plus terminal. Note that the two ac terminals are labeled with a sine wave symbol, and the + and − terminals are the rectified output.

SUMMARY OF DESIGN PROCEDURE

Now that you have a good idea of how half-wave, full-wave, and bridge-type power supplies work, let's summarize the key points in designing a power supply and choosing the parts to build one. As the designer, you must first consider two main specifications: dc output voltage V_o and dc output current I_o.

Remember, for low-current power supplies (say less than 25 mA), a half-wave rectifier circuit will be adequate, but for larger output currents you should use a full-wave circuit; otherwise the filter capacitor will be excessively large.

There are three major components to choose for any supply; the transformer, the diodes, and the capacitor. Having decided on the output voltage and current of the supply, you can choose a transformer. Recall that, for a half-wave or full-wave bridge circuit, the output voltage is approximately equal to the peak voltage of the secondary. It will be slightly less than this if you consider the drop across the conducting diodes and the ripple voltage, but generally these are very small; therefore, choose a transformer whose *peak* secondary voltage is about equal to, or slightly larger than, the dc output voltage you want, and one having sufficient current capability. The transformer secondary voltages are given in rms values. So you must first convert the peak value to rms by $V_{rms} = 0.707\ V_p$, then find a standard transformer with ratings close to what you need. Manufacturers also list a few other characteristics about transformers, such as price, physical dimensions, and insulation. The insulation rating tells you how much voltage the transformer can withstand between windings, or between case and one winding, without breaking down. Once you have found a few transformers with the current and voltage ratings you need, you can look at the other characteristics and make a final choice.

Next, choose the proper diodes. The diodes must have a forward current I_f equal to or greater than the maximum current you expect to draw from the supply. You may want to use separate diodes, or perhaps a rectifier assembly if you are building a bridge-type supply. Be sure to examine the prv rating of the diodes. Manufacturers will also list such things as price, case type, and operating temperature, which will probably be of interest.

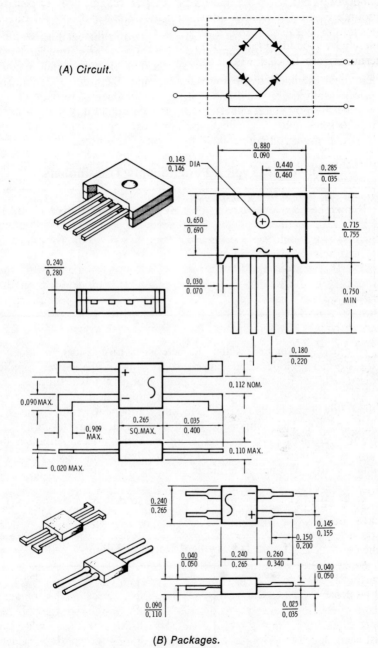

(A) Circuit.

(B) Packages.

Fig. 1-16. Bridge rectifier assembly.

Finally, choose a capacitor with a working voltage at least equal to the dc output voltage, and a capacitance at least as great as the value you calculated from the formula given. Listed in the catalogs you will also find price, physical dimensions, operating temperature, and possibly some information about leakage current. Be sure to find equivalent devices made by a few different manufacturers and compare prices. Many times you'll find that some components are quite expensive because they are made to withstand unusually high temperatures or shock for military applications. For most applications, just get the least expensive components that meet your basic needs.

Table 1-2. Typical Power Supply Transformers

Type	Secondary Volts	Amps	Price
P 8389	6.3	1	$3.14
P 6456	6.3 ct	6	7.31
P 5016	10.0	4	9.97
P 8136	12.6 ct	2	4.84
P 8357	25.2 ct	2	6.56
F 45X	24 ct	1	3.77
F 90X	10-20 ct, 40 ct	0.1	4.33
F 91X	10-20 ct, 40 ct	0.3	5.12

Note: All transformers rated for 115 Vac rms primary, 60 Hz.

Table 1-1 summarizes the characteristics to consider in selecting transformers, diodes, and capacitors for any of the power supplies discussed. Use it as a guide in designing your own power supplies.

To test your understanding of power supply design, work through the following example designs, selecting components from Tables 1-2, 1-3, and 1-4. The components listed in the tables were taken directly from electronics parts catalogs along with the prices. Although the prices, of course, are subject to change, they should give you some comparative information to aid in selecting components.

EXAMPLE 1-2—Design a power supply to develop about 55 volts dc at 10 mA. The ripple voltage should be no more than 1 volt

Table 1-3. Typical Rectifier Diodes

Type	PRV (Volts)	I_r (Amps)	Temp (°C)	Price
1N440	100	.03	150	$ .75
1N1581	50	5.0	150	1.12
1N2070	400	.75	25	.41
1N4001	50	1.0	75	.38
1N4719	50	3.0	75	.66
1N5003	1000	3.0	75	3.35

Table 1-4. Electrolytic Capacitors

Type	C (μF)	V dcw	Price
TE 1133	50	12	$ .63
TE 1135	100	12	.84
TE 1140	390	12	.96
TVL 1220	500	25	1.77
TVL 1230	1000	25	2.46
18F2454	5500	25	2.80
18F2455	8200	25	3.23
18F2456	18,000	25	5.22
15F164	100	50	.81
15F197	250	50	1.05
15F167	100	150	105
15F118	250	150	1.65

peak-to-peak. Select the following components from tables in this chapter:
Transformer type _____
Diode type _____
Capacitor type _____

SOLUTION—Since only 10 mA will be drawn, a half-wave rectifier supply will be sufficient.
Transformer:

$$V_s = 0.707 V_o = 0.707 \times 55 = 38.9 \text{ Vrms}$$
$$I_s = 2.3 I_o = 2.3 \times 10 \text{ mA} = 23 \text{ mA}$$

From Table 1-2 we see that a type F90X has a secondary voltage of 40 V center tapped and 100 mA, so it would give us an output voltage of

$$V_o = 1.41 \times 40 = 56.5 \text{ volts}$$

This is very close to the required voltage. If necessary, a small resistor can be placed in series with the load to drop the load voltage to 55 volts. Transformer type: *F90X*.
Diodes:

$$I_f \geqq I_o \geqq 10 \text{ mA}$$
$$prv \geqq 2V_o \geqq 2 \times 56.5 = 113 \text{ V}$$

From Table 1-3, the most economical choice would be *IN2070*.
Capacitor:

$$C_1 \geqq \frac{It}{V_r} = \frac{1 \times 10^{-2} \times 1.67 \times 10^{-2}}{1} = 167 \text{ μF}$$

Vdcw $\geqq V_o \geqq 55$ V

Since V_o exceeds 50 volts here, the next higher value listed in Table 1-4 is a 150-volt unit. A good choice would be the *15F118*.
The complete supply is shown in Fig. 1-17.

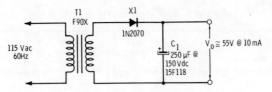

Fig. 1-17. Power supply for Example 1-2.

EXAMPLE 1-3—Design a power supply to deliver about 16 Vdc at 0.5 amp with a ripple of no more than 0.5 volt p-p.
  Transformer type _____
  Diode type _____
  Capacitor type _____

SOLUTION—
Transformer:

$$V_s \cong 0.707 V_o = 0.707 \times 16 = 11.3 \text{ Vrms}$$
$$I_s = 1.2 I_o \cong 1.2 \times 0.5 = 0.6 \text{ A}$$

Since the current required is relatively large, we should use a full-wave or bridge supply. A 12.6-volt filament transformer would be a good choice if we use a bridge, or a 24-volt ct transformer for a full-wave nonbridge supply. Due to its low cost, the *type F45X* transformer looks like a good choice for a full-wave nonbridge supply.

For the nonbridge supply, we can find the proper diodes by

$$I_f \cong 0.5 I_o \cong 0.5 \text{ amp}$$
$$\text{prv} \cong 2 V_o \cong 2 \times 16 = 32 \text{ volts}$$

From the table, type 1N4001 diodes are the best choice. Diode type *1N4001*.
  Capacitor:

$$C_1 \cong \frac{It}{V_r} = \frac{0.5 \times 8.35 \times 10^{-3}}{0.5} = 8350 \text{ } \mu F$$

$$V_{dcw} \cong V_o = 16 \text{ volts}$$

From the table, a *type 18F2455* has just about enough capacitance and the 25-volt rating is more than sufficient. If you want less ripple, you can go to the 18,000-μF unit, but the reduction in ripple might not be worth the added cost.

The complete supply is shown in Fig. 1-18.

Quiz

3. Design a full-wave bridge supply to deliver about 12 Vdc at 2.0 amps with no more than 2 V p-p ripple.
     Transformer type _____
     Diode type _____
     Capacitor type _____

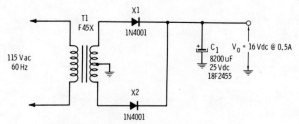

Fig. 1-18. Power supply for Example 1-3.

4. Design a bridge-type power supply to give about 16 Vdc output to a load that might draw up to 1.0 amp. V_r should be no more than 3 V p-p.

     Transformer type _____
     Diode type _____
     Capacitor type _____

 The components chosen as answers for these power supplies are not the only possible choices but merely good choices from the ones available. Whatever you design in electronics, remember that every design generally entails some compromise between characteristics, cost, size, availability of parts, etc. There is no single *best* design, although there may be several good designs based on many factors.

 In a later chapter we will discuss variable-output power supplies and regulated supplies. The supplies studied in this chapter will be the basic circuits on which we will expand.

2

Transistor Amplifier Design (Single-Stage)

An amplifier is basically a power converter, in that it converts dc power from the power supply into useful signal power. In this chapter we will examine single-stage (one-transistor) amplifiers and see how they are used to build up the power of small signals, such as those obtained from a phono pickup, microphone, photocell, etc. Since the signals obtained from such devices are usually small, on the order of millivolts, these amplifiers are called *small-signal* amplifiers, or *voltage* amplifiers. In a later chapter we will discuss *large-signal* (or *power*) amplifiers, such as those used, for example, to drive speakers.

BIPOLAR TRANSISTORS

The *bipolar* (or junction) transistor is basically a "sandwich" of different semiconductor materials in either of two forms, npn or pnp, as shown in Fig. 2-1A. The three parts of the transistor are called the *emitter, base,* and *collector.* Although modern fabrication techniques use somewhat different geometries, the operation is still the same, so we will use the "sandwich" model to illustrate the operation. The npn and pnp forms of the transistor operate almost identically, the only difference being the polarity of the power supplies used. Fig. 2-1B shows the schematic symbols for both types; Fig. 2-1C shows the bottom views of TO-18 and TO-5 transistor packages which are common formats for either type of transistor.

Transistor identification numbers begin with 2N (e.g., 2N1304, 2N2404, etc.). For the circuits discussed in this chapter, you can use a general-purpose or audio-frequency transistor, obtainable from any electronics parts store. The ratings are not critical as long as the tran-

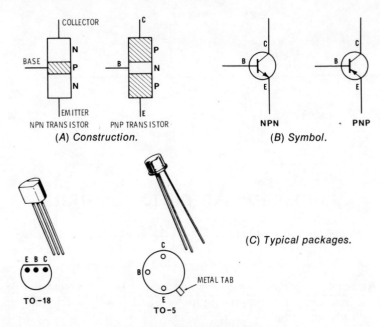

Fig. 2-1. Bipolar transistors.

sistor can withstand your power supply voltage. We will limit the discussion here to audio-frequency amplifiers, although most of the techniques can also be used at much higher frequencies.

As you will remember from the previous chapter on power supplies, a solid-state diode is a p-n junction. Looking at the npn transistor of Fig. 2-1A, you will notice that the base and emitter sections together form a p-n junction. Similarly, the collector and base sections also form a p-n junction. The two junctions in the transistor act like diodes connected back-to-back, as shown in Fig. 2-2A. Consequently, an ohmmeter connected between the base and emitter terminals (Figs. 2-2B and C) reads a low resistance in one direction and a high resistance in the other, just as for the ordinary diode. (Refer again to Chapter 1 if this procedure is not clear.) Also, an ohmmeter connected between the collector and base terminals again measures "diode action." Finally, connecting an ohmmeter between the emitter and collector with either polarity measures the high resistance of two diodes connected back-to-back, since one of the junctions is always reverse-biased. Sometimes a germanium transistor gives a lower resistance between emitter and collector if the negative lead of the meter is at the emitter. This is because the transistor is turned on a little with the leakage current generated by heat; we will say more about leakage current later.

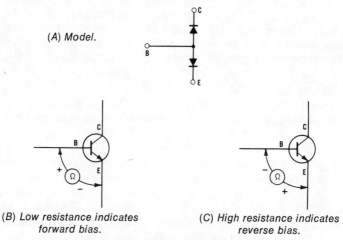

(A) Model.

(B) Low resistance indicates forward bias.

(C) High resistance indicates reverse bias.

Fig. 2-2. Checking an npn transistor.

The preceding measurements provide a quick way of checking whether a transistor is good or bad. If you measure diode action between the base and emitter terminals, and again between the base and collector terminals, and a high resistance between the emitter and collector, your transistor is probably good; otherwise, it is not. Also, these measurements will tell you quickly whether a transistor is npn or pnp. Try these measurements on several transistors yourself to see that you understand the method.

Quiz

1. Assume that your ohmmeter has a negative common. With the common connected to the base of an npn transistor, you measure a high resistance to the emitter and a low resistance to the collector. Is the transistor good or bad? Why?

2. With the common connected to the base of an unknown type of transistor, you read a low resistance to both emitter and collector. You then connect the common to the emitter and read a high resistance to both base and collector. Finally, you connect the common to the collector and measure a high resistance to both base and emitter. Is this transistor good or bad? Is it npn or pnp?

BIASING

Since an amplifier converts dc power into ac power, we can have both direct and alternating currents flowing through it. The direct current which flows is called the *bias* current. In order to have the amplifier operate correctly without distorting the signal, we must first bias the circuit properly. Biasing sets the operating point for the circuit and is similar to setting the idle on a car engine. Too much bias current wastes power and may also distort; not enough bias will cause the circuit to cut off (stall) momentarily and cause distortion. We will now take a look at the factors that affect the values of direct current flowing through a transistor.

The basic idea of current flow in a transistor is that the amount of current flowing from emitter to collector is controlled by the base current. But the base current is usually very small compared to the collector current. Thus, increasing or decreasing the small base current causes the much larger collector current to increase and decrease. The solid-state theory of current flow in a transistor is beyond the scope of this book. Many good physics texts provide explanations of what happens on the molecular level. As the circuit designer, however, you need not be familiar with solid-state physics in order to control the operation of transistors. Just be aware that ordinary transistors are current-controlled devices; that is, a small change in input current generates a large change in output current.

Take a look at the circuit of Fig. 2-3A. Note that the collector is connected through resistor R_C to the positive terminal of the V_{CC} supply. This is the normal polarity for an npn transistor. The collector-base diode is reverse-biased (Fig. 2-3B), so no current will flow through R_C as long as the base lead is open. But if you forward-bias the base-emitter diode as in Fig. 2-3C, base current will flow. At this time the much larger collector current will also flow. The exact relationship between the collector current and base current represents the current gain, called β (beta) or h_{FE}, and is mathematically expressed as

$$h_{FE} = \beta = \frac{I_C}{I_B} \qquad \text{(Eq. 2-1)}$$

The value of h_{FE} is given in manufacturers' spec sheets, typically ranging from about 20 to several hundred for various transistors. Thus for a transistor with a β of 50, the collector current is 50 times as large as the base current.

The arrows in Fig. 2-3D show the direction of electron flow in a transistor when the base is forward-biased. Notice that the emitter current (I_E) is the sum of I_B and I_C. This is always true in a transistor, but since the value of I_B is usually very small compared to I_C (by a factor of β), we can say that $I_C \cong I_E$.

Now let's make a simplification in the circuit of Fig. 2-3D. Note that both the V_{BB} supply and the V_{CC} supply are positive with respect to ground. We can then eliminate the V_{BB} supply merely by connecting one end of R_B to the V_{CC} supply as shown in Fig. 2-4. We will

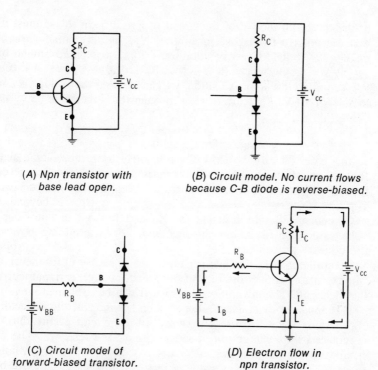

(A) Npn transistor with base lead open.

(B) Circuit model. No current flows because C-B diode is reverse-biased.

(C) Circuit model of forward-biased transistor.

(D) Electron flow in npn transistor.

Fig. 2-3. Collector current flows only when base current flows.

use this simpler, but equivalent, circuit of Fig. 2-4, called a *base-biased* circuit, to study biasing. Rearranging equation 2-1, we see that $I_C = \beta\, I_B$.

Values of β for each type of transistor are given by the manufacturer, but the actual values vary for different transistors of the same type. For example, one manufacturer gives β for a 2N1304 as having a minimum value of 40 and a maximum value of 200. If you want to know the exact value of β for any particular transistor, it can be easily

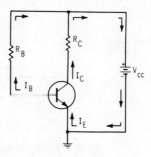

Fig. 2-4. Base-biased circuit.

33

determined with the aid of the circuit of Fig. 2-5. In this circuit, note that the voltage from base to emitter, V_{BE}, is very small (approximately 0.3 volts for germanium transistors) compared to the V_{CC} supply voltage. So the voltage across R_B is $V_{CC} - V_{BE} = 10 - 0.3 = 9.7$ V. For a close approximation, we can say that the voltage is equal to V_{CC}, or 10 V. The current (I_B) through R_B is therefore

$$I_B \cong \frac{V_{CC}}{R_B} \qquad \text{(Eq. 2-2)}$$

In this case, $I_B \cong 10$ V$/1$M $= 10$ μA. Measuring the voltage across R_C, you can determine the collector current. Suppose this is 1.5 volts; then $I_C = 1.5$ V$/1$K $= 1.5$ mA, from which $\beta = I_C/I_B = 1.5$ mA$/10\mu$A $= 150$.

Of course different resistors for R_B and R_C and different supply voltages can be used, but the technique for determining β is still the same.

Determining β in this manner turns out to be useful whenever you want to match transistors or find out which of several transistors has the highest β. The advantage of a high β will be explained later.

It is apparent that, since I_C depends on I_B, and since I_B depends on R_B, varying R_B will cause I_C to vary. In the circuit of Fig. 2-6, R_B is replaced with a large pot in series with a fixed resistor. As the resistance of the pot is increased or decreased, I_B will vary, causing I_C to vary. The fixed 20K resistor is in the circuit as a safety device; if R_B gets too small, I_B might become large enough to damage the transistor. Typical values for I_B in a circuit like this should be less than a few hundred microamps.

The best way to understand the following discussion is to build the circuit of Fig. 2-6 and experiment with it yourself using any available npn transistor. You can also use any value of supply voltage from a few volts up to at least 20 V or so. If in doubt, always consult a transistor manual to find the maximum voltage the transistor can tolerate.

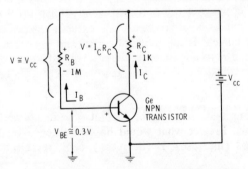

Fig. 2-5. Voltage drops in base-biased circuit.

If you connect a dc voltmeter from the emitter (ground) to the collector, the voltage V_C will be the difference between the supply voltage and the drop across R_C caused by I_C, or,

$$V_C = V_{CC} - I_C R_C \qquad \text{(Eq. 2-3)}$$

Now vary the base current and watch V_C. First, if the base lead is open, $I_B = 0$, so $I_C \cong 0$; in other words, the transistor is cut off. This is one extreme condition to avoid in building an amplifier. Next, if you connect the resistor to the base so that some base current flows, some collector current will also flow. As you decrease R_B, the collector current increases and V_C decreases. But as you continue to decrease R_B you eventually reach a point where V_C approaches zero. This means that all of the power supply voltage is dropped across R_C. This condition is called saturation and is the other extreme to be avoided in an amplifier.

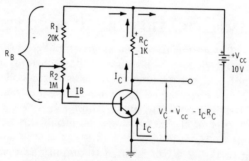

Fig. 2-6. Variable R in base lead varies I_C.

Here is another interesting point to consider. Say you have R_B adjusted so that $V_C = 9$ V when $V_{CC} = 10$ V and $R_C = 1$ K. The collector current must then be 1 mA since it causes 1 V to be dropped across R_C. Now without changing R_B, replace R_C with a 2K resistor. You will find that V_C decreases to 8 V; 2 V have been dropped across R_C. Finally, if you change R_C to 5K there will be 5 v dropped across it, showing that I_C is still 1 mA. This experiment demonstrates that the collector current is independent of R_C and V_C—that is, the transistor acts as a *constant current source*. This is a very useful characteristic of the transistor which will be brought up again later.

Quiz

To test your understanding of current flow in a transistor, try the following quiz. Predict what would happen to each of the following by answering I (increase), D (decrease), or S (remain the same).

Suppose at the start that R_2 in Fig. 2-6 is adjusted so that $V_C = 5$ V.

35

3. If R_2 is decreased by 50%,
    - (A) I_B will _____.
    - (B) I_C will _____.
    - (C) V_C will _____.

4. If R_C is decreased by 50% (V_C initially 5 V),
    - (A) I_B will _____.
    - (B) I_C will _____.
    - (C) V_C will _____.

5. Finally, if the circuit was originally constructed with $V_C = 5$ V and the transistor is then replaced with a transistor having a β 50% less than the original transistor,
    - (A) I_B will _____.
    - (B) I_C will _____.
    - (C) V_C will _____.

AMPLIFICATION OF AC SIGNALS

Now let's try using the transistor as an amplifier. For most applications, the actual values of bias current, base current, and collector current are not critical. As long as there is some collector current flowing, say 1 mA or so, and a few volts across the transistor, it will operate. But since you want to avoid cutoff ($V_C = V_{CC}$) and saturation ($V_C = 0$), a good choice for V_C is about half way between cutoff and saturation, or $V_C \cong \frac{1}{2} V_{CC}$.

The circuit of Fig. 2-6 can be used to amplify ac signals. All you have to do is feed a signal to the base with respect to ground through a capacitor, as shown in Fig. 2-7, and an amplified signal will appear between the collector and ground. The ratio of output voltage to input voltage is called the voltage gain, A_v, of the circuit, or

$$A_v = \frac{v_o}{v_{in}}$$

Since this circuit has the emitter grounded, it is called a *grounded-emitter,* or *common-emitter,* amplifier. The value of coupling capacitor C_c is not critical, just so it is large enough to have a small impedance, say about 100 ohms or less, at the lowest frequency to be amplified. Also, you can use any audio frequency to test the circuit. The oscillator of Fig. 7-3 in Chapter 7 will work well for such testing.

If possible, build the circuit of Fig. 2-7 and perform the following tests. If you cannot build it, imagine you are testing it. Say you are feeding an ac signal into the input of the circuit and R_B is adjusted so that $V_C \cong \frac{1}{2} V_{CC}$. Adjust the input signal amplitude so that the ac output signal is about 2 V p-p. Now with either a dc scope, or a dc voltmeter and an ac scope, observe the output at the collector. You will see that initially the collector voltage looks like Fig. 2-8A. Now connect the scope from base to ground. You should see a very small signal, on the order of less than 100 mV, showing that the circuit is amplifying; that is, v_o is much greater than v_{in}.

Next, connect the scope back to the collector and gradually reduce the value of R_2. Since I_B increases, I_C also increases and V_C decreases; the ac signal will be riding on a lower dc level, as in Fig. 2-8B. Finally, if you reduce R_2 far enough, the output will begin to get clipped on the negative peaks when V_C gets too low as shown in Fig. 2-8C. This distortion is due to the momentary saturation of the transistor on negative peaks. Next, if you increase R_B to too large a value, the transistor will be operating with too high a value of V_C (or I_C too low), thus causing clipping on the positive peaks whenever the transistor is driven into momentary cutoff (Fig. 2-8D). However, if the transistor has a very high β, you may not be able to decrease the base current sufficiently to reach cutoff. You can further decrease I_B by putting in a large fixed resistor, say 1M, in place of R_1.

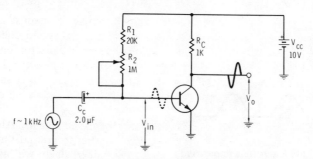

Fig. 2-7. Common-emitter amplifier.

From the above tests you can see that a good operating point is somewhere between cutoff and saturation. A value of $V_C \cong \frac{1}{2} V_{CC}$ will allow you to get the largest possible unclipped signal output from this circuit.

Now we can take a look at how to vary the amplifier gain. Suppose you have adjusted R_B so that $V_C = 8$ V, and the input signal is adjusted for an ac signal output of 2 V p-p. Now replace R_C with a 2K resistor and you will find that the ac output signal just about doubles. If you change R_C to a 3K, the gain will be about 3 times as great as in the original circuit. From this we can conclude that *the gain of the amplifier varies directly with the value of resistance in the collector.* This is actually to be expected, since you have already found out that the transistor acts like a constant current source. That is, regardless of the value of R_C, the current through it, ac or dc, will remain constant as long as the transistor operates between cutoff and saturation.

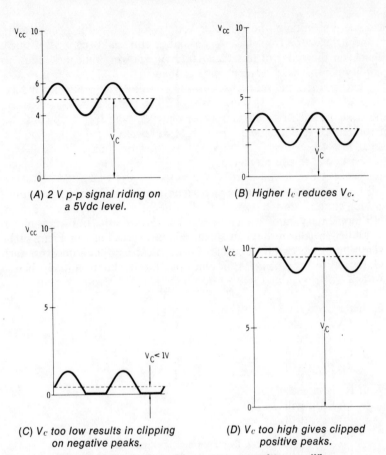

Fig. 2-8. Collector output from common-emitter amplifier.

So the larger R_C is, the greater the signal voltage drop across it; hence, the larger the output signal voltage.

Usually, the output from a single transistor will not give sufficient amplification; in any case, you will want to couple the ac signal from the collector to some other circuit or to a load. We will use resistor R_L in Fig. 2-9 to represent this additional ac load on the stage. Suppose you originally set V_C to 5 V with pot R_2. Before connecting R_L to the collector, adjust the input signal so that $v_o = 2$ V p-p. Now connect R_L (about 1K). You will see that v_o decreases. The reason is that the *ac load* seen by the collector has increased. Let's use the symbol r_L to represent the ac load (the parallel combination or R_C and R_L) seen by the collector. Try changing R_L higher and lower, and you will find that the gain varies directly with total r_L. Capacitor C_2 is used between R_L and the collector of the transistor to block dc from the load.

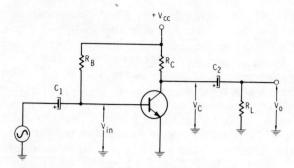

Fig. 2-9. R_L acts as a partial load on collector output.

Quiz

Predict what will happen to the following readings in Fig. 2-9 by choosing I (increase), D (decrease), or S (remain the same) for each of the following changes. Assume that the circuit is initially biased with $V_C = \frac{1}{2} V_{CC}$, and that v_o is initially about 2 V p-p.

6. If R_C is decreased by 50%,
    (A) v_{in} will _____.
    (B) V_C will _____.
    (C) A_v will _____.
    (D) v_o will _____.

7. If R_C is increased by 50%,
    (A) v_{in} will _____.
    (B) V_C will _____.
    (C) A_v will _____.
    (D) v_o will _____.

8. If R_L is decreased by 50%,
    (A) v_{in} will _____.
    (B) V_C will _____.
    (C) A_v will _____.
    (D) v_o will _____.

9. If R_L is increased by 50%,
    (A) v_{in} will _____.
    (B) V_C will _____.
    (C) A_v will _____.
    (D) v_o will _____.

VOLTAGE DIVIDER BIAS

In a previous section you saw how I_C depends on β. Suppose you build an amplifier using the circuit of Fig. 2-9, and R_B is selected to make $V_C = \frac{1}{2} V_{CC}$. Assume that you are using a transistor with a β of 50, and you do not want to use a pot for R_B in the final circuit. Certainly you could set the operating point with a pot, remove the pot from the circuit, measure its resistance with an ohmmeter, and replace it with the closest standard fixed resistor. This would work perfectly well. However, suppose after some time, the transistor became accidentally damaged and you had to replace it. What do you suppose would happen if you replaced it with another transistor with a β of 100 or more? I_B would remain the same as before; that is, $I_B = V_{CC}/R_B$, but the collector current would be at least twice as great as before. This, of course, means that the transistor would go into saturation. You could solve this problem in a couple of different ways. You could either measure several transistors and find one with a β close to the one you are replacing, or you could re-adjust R_B. But you can see that for a manufacturer going into production of the amplifier, measuring the β of each transistor, or using different values of R_B for each amplifier would be extremely costly. Obviously, a better way would be to use a biasing method in which the collector current is independent of β.

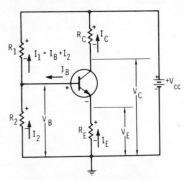

Fig. 2-10. Voltage-divider biasing.

The biasing arrangement of Fig. 2-10 makes the collector current essentially independent of β. That is, for almost any transistor you use in the circuit, the collector current will be the same value and will be determined only by the resistors in the circuit.

Here's how the circuit works. Resistors R_1 and R_2 form a voltage divider across the V_{CC} supply. The direction of electron flow is shown in the figure. Notice that $I_B + I_2 = I_1$. If the resistors chosen are small enough so that I_2 is much greater than I_B, then $I_2 \cong I_1$. The reason for this approximation is simply that if we make I_2 large enough compared to I_B, we can ignore I_B and say that the voltage appearing between the base and ground, V_B, is determined only by resistors R_1 and R_2. For example, if $R_1 = 40K$, $R_2 = 10K$, and $V_{CC} = 10$ V, the current through R_2 and R_1 would be $I = 10$ V/50K $= 200$ μA; so

$V_B \cong 200 \ \mu A \times 10K = 2$ V. This would be approximately true even if the base current of the transistor is 10, 15, or 20 μA.

Assume we are using a germanium transistor where the normal dc voltage drop from base to emitter is only about 0.3 V, regardless of β. That means that the voltage V_E, measured between emitter and ground, is almost equal to V_B, or just slightly less than 2 volts. Mathematically, the relationship is governed by

$$V_E \cong V_B \cong V_{CC} \times \frac{R_2}{R_1 + R_2}$$

And the current through R_E, which is I_E, can be found by

$$I_E = \frac{V_E}{R_E}$$

Suppose you make the emitter resistor 2K; then $I_E \cong 2$ V/2K $= 1$ mA. Since the collector current is virtually equal to the emitter current, $I_C \cong 1$ mA. Notice that we are able to determine the collector current without knowing β for the transistor; it doesn't really make any difference what the value of β is for the transistor, since the collector current is controlled entirely by resistors R_1, R_2, and R_E. Consequently, almost any transistor can be used in the circuit and not change its operation. As usual, the transistor must not be saturated in order for it to operate properly.

EXAMPLE 2-1—Suppose that in Fig. 2-10, $R_1 = 60K$, $R_2 = 20K$, $R_E = 2K$, and $V_{CC} = 16$ V. Assuming the transistor is not saturated, what is the collector current?

SOLUTION—First of all,

$$V_B \cong V_E \cong 16 \left(\frac{20K}{20K + 60K} \right) = 4 \ V.$$

Then,

$$I_C \cong I_E \cong 4 \ V/2K = 2 \ mA.$$

The voltage at the collector with respect to ground can again be found by equation 2-2, but the voltage between collector and emitter will be less than V_C due to the drop across R_E. You must make sure that there are at least a few volts between emitter and collector in order not to saturate the transistor.

When amplifying ac signals, we must connect a capacitor from emitter to ground, as in Fig. 2-11, to make the emitter grounded to ac. This capacitor should have a fairly high capacitance, usually 10 μF or larger for audio work. In some circuits, the capacitor is omitted to give the circuit more gain stability, but with a reduction in maximum gain. This subject will be treated in the next chapter.

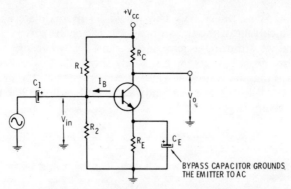

Fig. 2-11. Grounded-emitter amplifier using voltage-divider bias.

Now that you have a general idea of how the circuit is biased, let's set up a "thumb rule" design for the amplifier of Fig. 2-11. First of all, we must avoid clipping. The larger the dc voltage between collector and emitter, the larger the output voltage can swing before clipping occurs on the negative peaks. On the other hand, we also want an equally large dc voltage across R_C, so as not to clip the output on positive peaks. Any dc voltage dropped across R_E is wasted as far as the signal is concerned, because the emitter is held at ac ground by the bypass capacitor C_E. However, we do need some dc voltage drop across the emitter resistor in order to control the bias current. If we use only a small part of the power supply voltage (say 10%) across R_E, most of it will still be left for signal purposes; so assume $1/10$ of the V_{CC} supply is dropped across R_E.

Now what current should we use? In small-signal amplifiers, a good arbitrary value to use for I_C is about 1 mA. With 1 mA flowing, the transistor will operate well but will not be wasting power. So, the first thing to do is choose R_E to make $I_C \cong 1$ mA with $1/10$ of V_{CC} dropped across it.

Next determine the values of voltage-divider resistors R_1 and R_2. If $1/10$ of the V_{CC} supply appears across R_E it must also appear across R_2; Resistor R_2 should therefore be $1/10$ of the total resistance $R_1 + R_2$, or in other words, $R_1 = 9R_2$. A good value of R_2 is about 10 times R_E. This will further ensure that I_B is negligible compared to I_2.

Finally, we must choose a value for R_C. If you do not have to work around some specific value of R_C, such as the impedance of headphones or the like, you can use almost any value. Remember, the larger R_C, the higher the voltage gain; but if R_C is too large, the circuit may clip on negative peaks. A good choice for R_C is probably one that will drop about half of the remaining power supply voltage with I_C flowing through it. In the next chapter we will take a look at some of the other factors that affect voltage gain, but, generally, the larger R_C is, the greater the voltage gain will be.

To summarize the thumb-rule design:

1. Choose I_C (say 1 mA).
2. Choose R_E to drop 1/10 of V_{CC} with $I_C = I_E$ flowing through it.
3. Choose $R_2 = 10R_E$.
4. Choose $R_1 = 9R_2$.
5. Choose R_C to drop half of the remaining power supply voltage with I_C flowing through it.

EXAMPLE 2-2—In the circuit of Fig. 2-11, suppose you are using a V_{CC} supply of 18 volts. Design the amplifier circuit assuming $I_C = 1$ mA.

SOLUTION—

$$R_E = \frac{V_E}{I_E} = \frac{\frac{1}{10} V_{CC}}{I_E} = \frac{1.8 \text{ V}}{1 \text{ mA}} = 1.8 \text{ K}$$

$$R_2 = 10 \; R_E = 18 \text{ K}$$

$$R_1 = 9 \; R_2 = 162 \text{ K}$$

$$R_C = \frac{\frac{1}{2}(V_{CC} - V_E)}{I_C} = \frac{\frac{1}{2}(18 - 1.8)}{1 \text{ mA}} = 8.1 \text{ K}$$

In the final circuit, choose the closest standard resistor values. If V_C must be some exact value, you can put in a pot for R_1 and make it adjustable. Also, after setting the bias for one transistor, replace the transistor with several others, one at a time, and see how the collector current and voltage remain essentially constant. Here is the primary advantage of this method of biasing over the simple base-biased circuit.

Quiz

Suppose you have built the circuit as described in the previous example. Predict what will happen in each of the following cases by choosing I (increase), D (decrease), or S (remain the same). (Choose I or D only if the change would be quite noticeable.)

10. If R_1 is increased 50%,
    (A) V_B will _____.
    (B) V_E will _____.
    (C) I_C will _____.
    (D) V_C will _____.

11. If R_2 is increased 50%,
    (A) V_B will _____.
    (B) V_E will _____.

(C) I_C will _____.
   (D) V_C will _____.
12. If R_E is increased 50%,
   (A) V_B will _____.
   (B) V_E will _____.
   (C) I_C will _____.
   (D) V_C will _____.
13. If R_C is increased 50%,
   (A) V_B will _____.
   (B) V_E will _____.
   (C) I_C will _____.
   (D) V_C will _____.
14. If the transistor is replaced with a different one whose β is 50% higher than that of the original transistor,
   (A) V_B will _____.
   (B) V_E will _____.
   (C) I_C will _____.
   (D) V_C will _____.

TEMPERATURE EFFECTS

The circuit of Fig. 2-11 is better than the base-biased circuit because the collector current is almost independent of β. One transistor can be replaced with another without seriously changing the bias current. It has another advantage, too: it is more temperature stable.

In the simple base-biased circuit of Fig. 2-5, you would find that the collector current increases drastically with an increase in temperature. To demonstrate this, try building the circuit with a germanium transistor and set $V_C = \frac{1}{2}V_{CC}$. Then hold a low-wattage soldering iron against the transistor for about 5 seconds while watching the collector voltage V_C with a meter. You will see that V_C decreases, even to the point of saturation in some transistors. (This is far more noticeable with germanium transistors than with silicon.) The reason behind this increase in "leakage current" is not important here, but the fact that it increases with temperature *is* important.

Now if you build the circuit of Fig. 2-10 with the same transistor used in the circuit of Fig. 2-5, you will find that the collector current increases only slightly under the same test conditions (soldering iron for 5 seconds). This shows that the circuit of Fig. 2-10 is far more temperature-stable than the base-biased circuit. The stabilizing component is emitter resistor R_E. As the collector current starts to increase, due to an increase in temperature, the voltage across R_E starts to rise. This voltage, V_E, bucks the base voltage and tends to turn the transistor off slightly. The net effect is less increase in I_C.

There are several other methods of biasing transistor circuits, but if you understand the factors that govern current flow in the two circuits discussed, you should already be able to figure out some others.

For example, in Fig. 2-12 there are three common-emitter amplifier circuits using various biasing arrangements. Try building these circuits and experiment with various resistors to see if you understand how to control the collector current and voltage. Remember, anything that increases I_B also increases I_C. Try feeding an ac signal in, to measure the amount of amplification, and try the temperature stability test. You should always use a fixed resistor (not shown) in series with R_B in each of the circuits, to limit the base current and avoid possible damage to the transistor.

SUMMARY

In this chapter we studied the common-emitter amplifier. We saw that the transistor must be biased properly in order to amplify. We

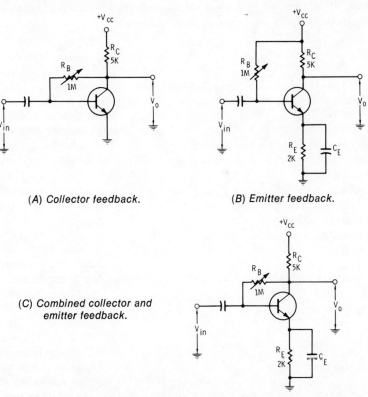

(A) Collector feedback.

(B) Emitter feedback.

(C) Combined collector and emitter feedback.

Fig. 2-12. Three biasing arrangements for a grounded-emitter amplifier.

also saw that the collector current is controlled by the base current as well as by the β of the transistor, or in general, $I_C = \beta I_B$. We looked at the base-biased circuit of Fig. 2-4, and found that $I_B \cong V_{CC}/R_B$, and that $V_C = V_{CC} - I_C R_C$.

We also found that increasing the collector load resistance increases the voltage gain of the circuit. The complete base-biased amplifier is shown in Fig. 2-9.

Next we examined a better method of biasing the common-emitter amplifier, in which the collector current is largely independent of β. This circuit (Fig. 2-11) uses a voltage divider and emitter resistor to establish the bias point. It was seen to be much more temperature stable than the base-biased circuit. A "thumb-rule" design for the amplifier was formulated and summarized.

Remember that frequently the *best* way to understand transistors is to build circuits and vary some of the component values while observing what happens in various parts of the circuit. One rule to keep in mind is to be systematic and *make only one change at a time;* otherwise, it is easy to lose track of the effects that each change has produced.

3

Transistor Amplifier Design (Cascaded Stages)

In the previous chapter, we saw that connecting an additional ac load (r_L) across the output of an amplifier stage reduces the gain of the stage. This load (r_L) is usually not just a resistor, but represents the input impedance (ac resistance) of the following stage.

INPUT IMPEDANCE

Fig. 3-1 shows a two-stage transistor amplifier. The first stage is a typical amplifier like the one studied in Chapter 2. Capacitor C2 couples the ac signal from the collector of Q1 to the input of Q2. This type of coupling is called RC coupling. In Fig. 3-1B we see the ac equivalent circuit of Q1 coupled to Q2. The equivalent ac resistance (r_L) seen by the collector of Q1 is the parallel combination of R1, R2, R3, and r'_{in}, where r'_{in} is the equivalent ac input resistance seen looking into the base of Q2. Notice that the ac load presented by coupling capacitor C2 is assumed to be negligible. It is evident that the ac input resistance of Q2 will significantly affect the gain of Q1; the lower r'_{in}, the lower the gain of Q1, etc. In order to work effectively with the cascaded amplifier, we would like to know what factors affect this input impedance.

Without going through a formal, mathematical model, we can set up a simple circuit to demonstrate what factors affect the input resistance. You must remember that we are talking about the *impedance*, or ac resistance, seen looking into the base of the transistor stage. The ac resistance cannot be measured with an ohmmeter, but must be obtained using an ac signal. With the test circuit of Fig. 3-2, we can also learn more about factors that affect the voltage gain of the circuit.

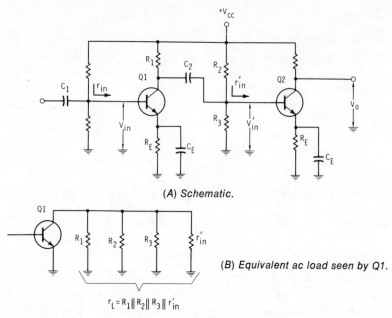

Fig. 3-1. Two-stage cascaded amplifier.

To the right of point B in Fig. 3-2 is the typical voltage amplifier circuit studied in Chapter 2. Notice that the dc voltage at point B with respect to ground is about 2 V; likewise the voltage from emitter to ground is near 2 V. So the emitter current is approximately 2 V/2K = 1 mA.

Now examine the circuit to the left of point B. First of all, we have a 100:1 voltage divider, consisting of R_A and R_B across the signal generator. This serves two purposes. One, if we do not have an instrument capable of measuring very small signals (on the order of a few millivolts) we still know that the signal amplitude (v_s) across R_B will be about 1/100 of the total voltage applied to R_A and R_B in series. We can easily measure this larger applied voltage with conventional meters. In this figure, the generator voltage is set at 1 V p-p, which means that v_s will be approximately 10 mV p-p.

Two, since the signal across R_B is used as the source driving the amplifier, the whole divider acts like a signal generator with a very low internal impedance (about 10 ohms). The advantage here is that when the input impedance to the amplifier is connected across R_B, there will be practically no change in the voltage across R_B. In other words, there will be negligible loading effect on the signal source, as long as the input impedance to the amplifier is much higher than 10 ohms, which it will be in most cases.

To measure the input impedance to the amplifier, proceed as follows:

1. With R_X set to zero resistance, measure the ac output voltage v_o. (The circuit will be amplifying, so this output voltage can easily be measured. It will probably be on the order of about 1 to 2 V p-p.)
2. Next, start increasing R_X until v_o drops to half the value measured in step 1.
3. Finally, without changing the value of R_X, remove it from the circuit and measure it with an ohmmeter. The value of R_X you measure will be equal to the input impedance of the amplifier. Since the output dropped to half the previous value, and since nothing was done to the amplifier to change the gain, the input signal from point B to ground must have dropped to half its original value. The fact that equal voltages appear across R_X and the input to the amplifier means that R_X must be equal to the input impedance.

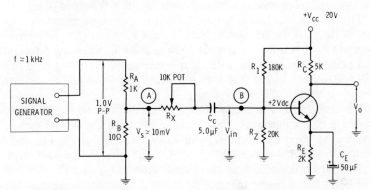

Fig. 3-2. Circuit for measuring input impedance of an amplifier stage.

Of course, if you have a very sensitive scope, you can measure v_{in} at point B directly, but measuring with the method described makes a sensitive scope unnecessary. In addition, the scope presents no loading effect this way, which it might when measuring across a very high impedance.

Now that you know how to measure input impedance, let's see what can cause it to vary.

Your measured value of input impedance to the stage was most likely much lower than just the parallel combination of R_1 and R_2. This indicates that r'_{in} (the resistance seen looking directly into the base of Q2) is the most significant resistance here. It turns out that the ac resistance, seen looking into the base when R_E is bypassed, varies inversely with the amount of emitter current flowing. A simple test will verify this.

In the circuit of Fig. 3-2, change R_E to 1K without changing any other values. I_E will increase to about 2 mA, since the voltage divider (R1, R2) still maintains about 2 V at the base, and hence across R_E. Now starting with R_X at zero resistance again, and $v_s = 10$ mV, measure v_o.

You should find that v_o is about twice as great as it was originally with $I_E = 1$ mA. This indicates that the gain of the circuit has just about doubled with a doubling of I_E. If you triple I_E, you find that the gain triples, etc., all of which means that *the gain varies directly with I_E*. This property will become very useful in later discussions.

Once again, to measure the input impedance use the procedure outlined before. Measure v_o while increasing R_X, until v_o drops to half its original value; then measure the input resistance. You should find that the input resistance to this stage has decreased to about half of what it was when I_E was 1 mA. In other words, *r'_{in} varies inversely with I_E*. Typical values of r'_{in} for a stage like that of Fig. 3-2 are in the range of from several hundred ohms to a few thousand ohms.

There is another factor besides I_E that affects input impedance, and that is the β of the transistor. You learned in the last chapter how to measure β. If you have a few different transistors, measure β for each and perform the input impedance measurements described here. You will find that *the higher β is, the larger the input resistance to the stage;* i.e., r'_{in} is directly proportional to β.

It was mentioned in the last chapter that there is some advantage to having a high value of β. Now you can see what the advantage is. If you are building an amplifier with cascaded stages as in Fig. 3-1A, using a transistor with a high β for Q2 will give the second stage a high input impedance. Thus the load resistance (r_L) on the first stage will be higher, causing the gain of the first stage to be greater. Generally, in circuits like that of Fig. 3-1, *higher β transistors will give a greater overall gain.*

If you have built the circuits and performed the tests described so far in this chapter, you have experimentally verified the following points:

1. The input impedance to a transistor stage decreases if I_E is increased.
2. The input impedance of a transistor stage is higher with higher values of β.
3. The gain of a transistor stage increases if I_E is increased.

For those who like to have a mathematical "handle" on circuits, we will now briefly discuss how to calculate the input impedance and the voltage gain.

It can be shown by a fairly rigorous analysis that a diode has an ac resistance at room temperature approximately equal to 25 mV/I, where I is the dc forward bias current through the diode in amps. That is, if the current through it is 1 mA, the ac resistance is about 25 mV/ 1 mA = 25 ohms.

Recall that the base-emitter section of a transistor is also a diode. However, when used as a grounded-emitter amplifier, the ac resistance (impedance) seen looking into the base is the impedance of the diode multiplied by the β of the transistor. That is,

$$r'_{in} \cong \beta r_e$$

where r_e is the ac resistance of the emitter-base diode and is found by the relationship

$$r_e \cong \frac{25 \text{ mV}}{I_E}.$$

EXAMPLE 3-1—In the circuit of Fig. 3-2, if I_E is 1 mA and β is 100, what is the input resistance seen looking into the base?

SOLUTION—
$$r_e \cong 25 \text{ mV}/I_E \cong 25 \text{ mV}/1 \text{ mA} \cong 25 \text{ ohms}$$
and
$$r'_{in} \cong \beta \, r_e \cong 100 \times 25 \cong 2500 \text{ ohms}$$

We find that increasing I_E increases the gain; since increasing I_E decreases r_e, we can also say that *decreasing r_e increases the gain*. You will remember from Chapter 2 that the gain of the amplifier is directly proportional to the ac load resistance (r_L) seen by the collector. We can now state the voltage gain as

$$A_v \cong \frac{r_L}{r_e}$$

EXAMPLE 3-2—Calculate the input impedance and gain of the circuit of Fig. 3-3.

SOLUTION—First of all, we must determine I_E. Due to the voltage divider (R_1, R_2), we will have about 1/10 of V_{CC} or 2.5 V at the base with respect to ground. The voltage at the emitter with respect to ground will also be about 2.5 V. This means $I_E \cong 2.5 \text{ V}/1K \cong 2.5$ mA. Next we find $r_e \cong 25 \text{ mV}/2.5 \text{ mA} \cong 10$ ohms. Then the input resistance seen looking into the base is

$$r'_{in} \cong \beta \, r_e \cong 80 \times 10 \cong 800 \text{ ohms.}$$

Of course the total input impedance seen looking into the stage is the parallel combination of R_1, R_2, and r'_{in}. In this case, however, it will still be approximately 800 ohms, since R_1 and R_2 are very large by comparison with r'_{in}.

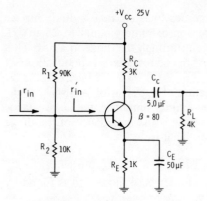

Fig. 3-3. Circuit for Example 3-2.

The voltage gain, A_v, is r_L/r_e, where r_L is the parallel combination of R_C and R_L. So,

$$A_v \cong \frac{1.71 \text{ K}}{10} \cong 171$$

If you are not interested in calculating the values of r_{in} and A_v, just remember the general rules listed earlier. By knowing what factors affect the gain and input impedance, you can work more effectively with existing amplifiers and modify them to meet your own requirements. In other words, when you want a circuit to perform a certain task, you don't have to "re-invent the wheel"; it is unnecessary to design every circuit from scratch. Most engineering design consists of taking standard circuits and modifying them to meet specific requirements. If you really understand the important factors governing the operation of a circuit, you can modify it to suit your particular needs.

Quiz

To see how well you understand the cascade transistor amplifier, try the following quiz. Answer I (increase), D (decrease), or S (remain the same) for these examples. Refer to Fig. 3-4. Assume that the circuit is working properly, that source voltage v_s remains constant at about 0.2 mV, and that v_o is originally about 2 V p-p.

1. If Q2 is replaced with a transistor having a β of 200, v_o will _____.
2. If R_3 is increased 20%, v_o will _____.
3. If R_4 is decreased 20%, v_o will _____.
4. If R_5 is decreased 20%, the gain of Q2 will _____.
5. If R_6 is decreased 20%, the gain of Q2 will _____.
6. If R_8 is decreased 20%, the gain of Q2 will _____.
7. If R_8 is decreased 50%, the ac load on Q1 will _____.
8. If R_8 is decreased 50%, the gain of Q1 will _____.

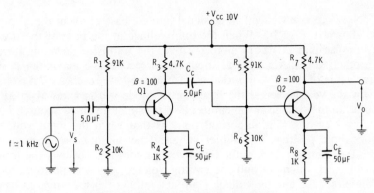

Fig. 3-4. Circuit for problem questions.

EMITTER FOLLOWERS

As you have learned, the gain of an amplifier varies directly with the load on it. It is then obvious that working into a load with very low resistance, say 50 ohms or less, you cannot get much gain. If you are familiar with transformers as impedance-matching devices, you will remember that a transformer can make a low value of resistance connected to the secondary side look like a high resistance at the primary. There is a way of using a transistor in a similar manner, called an *emitter-follower* or *common-collector* circuit.

Fig. 3-5 shows a typical emitter-follower circuit. Notice that the collector is connected directly to the power supply, so it is common to the ac signal. Notice also that the output is taken across the emitter resistor (R_E), which is not bypassed with a capacitor. Due to the voltage divider, R_1 and R_2, the voltage at the base with respect to ground is about 2 V, so the voltage at the emitter is also near 2 V (less the voltage drop across the base-emitter diode). Assuming a germanium transistor is used, the drop across the base-emitter diode is about 0.3 V, making $V_E \cong 1.7$ Vdc.

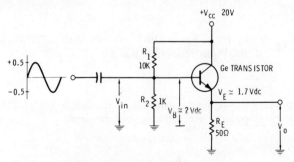

Fig. 3-5. Emitter-follower circuit.

Now suppose we feed an ac signal into the base of about 0.5 V p-p, as shown in Fig. 3-5. When the total voltage at the base of the transistor (ac plus dc) goes more positive, the transistor turns on more. In other words, when the ac input signal reaches the maximum positive value of +0.5 V, the total voltage at the base reaches +2.5 volts. Now the voltage across the base-emitter diode still remains about 0.3 volts, so the voltage at the emitter also rises half a volt to about 1.7 + 0.5 = 2.2 volts. The important point is that when the input voltage goes more positive, the emitter voltage (the output) also goes more positive. Similarly, when the input voltage goes less positive, the transistor turns off more and the emitter voltage goes less positive, too. Therefore, the emitter voltage (output) *follows* the variations in base voltage (input); thus, the name emitter follower is appropriate.

Since the voltage across the base-emitter diode remains essentially constant, the variation in output voltage is approximately the same as the variation in input voltage. That is, the amplitude of the output is the same as the amplitude of the input; *the gain of the circuit is unity,* and $V_o = V_{in}$.

You may wonder why we would use an amplifier with a gain of only unity, where there is no increase in output voltage; however, we do get *power* gain. That is, the signal power delivered to the load is greater than the input power delivered to the base. The reason is that the input impedance of the circuit is much higher than the load impedance; so if the same amplitude of signal voltage appears across the load as across the input, the load signal power will be greater than the input signal power.

You already know from your experience with dc biasing of transistors that the current flowing into the base is smaller than the emitter current by a factor of β (or $I_B = I_E/\beta$). The same is true for ac current as well. Since beta times as much current flows through R_E as flows into the base, the value of resistance in the emitter "appears" to be multiplied by beta when viewed from the base. We can state the input resistance r'_{in} mathematically as

$$r'_{in} \cong \beta R_E$$

This equation applies if R_E is much greater than r_e which is usually the case. If not, the input resistance will be

$$r'_{in} = \beta(R_E + r_e)$$

EXAMPLE 3-3—If the transistor in Fig. 3-5 has a β of 120, what is the input impedance seen looking into the base?

SOLUTION—
$$r'_{in} \cong \beta R_E \cong 120 \times 50 \cong 6K$$

So the transistor makes the 50-ohm load appear as a 6K resistance seen looking into the base. Thus the emitter follower does a job similar to a transformer in that it can make a low value of resistance look like a much higher value. More important, an emitter follower is usually smaller than a transformer and less expensive. Of course, the total resistance seen looking into the stage (to the right of the coupling capacitor) will be the parallel combination of R_1, R_2, and r'_{in}.

Fig. 3-6 illustrates one application of cascaded amplifiers. In Fig. 3-6A an amplifier stage is working directly into a small load impedance (50 ohms). Using the amplifier as shown would not give much gain because it is working into a low load resistance. But in the circuit of Fig. 3-6B, we have a second stage (Q2), which is an emitter-follower circuit coupled to Q1. The emitter follower makes load resistor R_L look like a much higher value to the collector of Q1, so the gain of the amplifier is greatly improved.

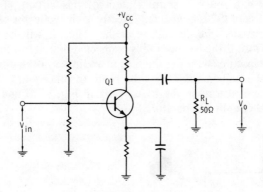

(A) Single stage has low gain.

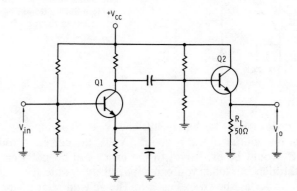

(B) Emitter follower increases gain of Q1.

Fig. 3-6. Emitter follower in cascaded amplifier.

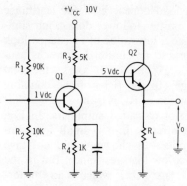
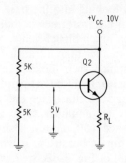

(A) Simplified bias arrangement.   (B) Equivalent bias at Q2.

Fig. 3-7. Eliminating bias network for Q2.

Next, we can make a simplification of this circuit (Fig. 3-7A), where the base of Q2 is connected directly to the collector of Q1. If Q1 is biased normally, the dc voltage at the collector will be about 5 V as shown. Thus, if the base of Q2 is connected directly to the collector, Q2 will be biased exactly the same as it would be in the circuit of Fig. 3-7B with the voltage divider at the base. In other words, we do not need the bias resistors on the base of Q2 in Fig. 3-6B and neither do we need the coupling capacitor.

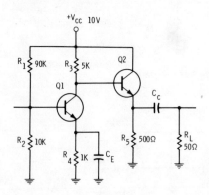

Fig. 3-8. Capacitive coupling eliminates dc emitter current through R_L.

The circuit of Fig. 3-7A may still have one disadvantage: since the voltage at the emitter of Q2 is about 5 volts, the current through load resistor R_L could be excessive. In the event you do not want any dc current through the load, you can capacitively couple the load to the emitter follower as shown in Fig. 3-8. The ac load seen by the emitter of Q2 is the parallel combination of R_5 and R_L. (Note that R_5 is the R_E for Q2.)

We can see that the ac resistance seen looking into the base of Q2 will be, in general:

$$r'_{in} \cong \beta r_L$$

where,
r_L is the parallel resistance of R_E and R_L.

Before finishing with emitter followers, we will look at one more arrangement that will make a small value of resistance look extremely large. The circuit shown in Fig. 3-9 is called a *Darlington* circuit.

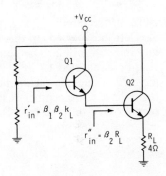

Fig. 3-9. Darlington circuit makes small load resistance look large at input.

Notice that it consists of two emitter followers in cascade. In this circuit, the ac resistance seen looking into the base of Q2 will be $r''_{in} \cong \beta_2 r_L$. The base of Q2 acts as the load on Q1, which is also an emitter follower, so the input resistance seen looking into the base of Q1 will be β_1 times the total load on Q1, or

$$r'_{in} \cong \beta_1 \beta_2 r_L$$

where,
r'_{in} is the effective input resistance of Q1,
β_1 is the β of Q1,
β_2 is the β of Q2,
r_L is the load resistance of Q2.

EXAMPLE 3-4—If both transistors have a beta of 100, what is the input resistance seen looking into the base of Q1 in Fig. 3-9?

SOLUTION—
$$r'_{in} \cong 100 \times 100 \times 4 = 40K$$

We will look at many examples of emitter followers and Darlington amplifiers in the chapters on power amplifiers and regulated power supplies. For the present, just remember that you can use an emitter follower any time you want to make a low value of resistance look like a much higher resistance.

FEEDBACK IN AMPLIFIERS

In the emitter-follower circuit, the input impedance is high because the emitter resistor is unbypassed. This technique can also be used in a common-emitter amplifier stage to increase the input impedance. At the same time, it gives a greater voltage gain than is possible with an emitter follower.

In the emitter-follower circuit of Fig. 3-10A, the input impedance seen looking into the base is approximately

$$r'_{in} = \beta \times R_E = 100 \times 1K = 100K$$

Next, in the circuit of Fig. 3-10B we see that the emitter resistor is also left unbypassed. In this circuit, the input impedance seen looking into the base will also be about 100K, even though the collector has a resistor connected to it. If possible, verify this for yourself experimentally using the techniques described earlier. If not, all that you really have to remember is that the input impedance to the circuit as seen looking into the base is essentially independent of whatever you

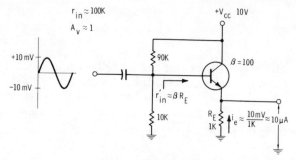

(A) Emitter-follower.

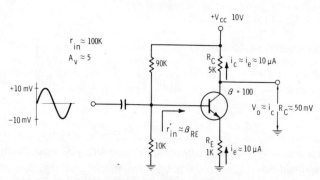

(B) Unbypassed CE amplifier.

Fig. 3-10. Comparing an emitter-follower circuit and an unbypassed common-emitter amplifier.

connect in the collector circuit (as long as the transistor is not saturated).

Now let's look at the amplification of the circuit. In the emitter-follower circuit of Fig. 3-10A, a 10-mV peak signal is applied to the base. Since the emitter-follower circuit has a gain of one, we know that the signal appearing at the emitter is also about 10 mV. Again, in the amplifier of Fig. 3-10B, we find that the signal appearing at the emitter with respect to ground is 10 mV, since this part of the circuit is exactly like the emitter-follower circuit. This 10-mV signal causes an ac emitter current of about 10 mV/1K ≅ 10 μA to flow through R_E. From previous work, you know that the ac collector current is about the same as the emitter current, so the ac collector current is also about 10 μA. This ac collector current flows through an R_C of 5K, causing an ac voltage to appear across it. The ac voltage appearing across this collector resistor can be used as the output voltage for the circuit. This output voltage for the circuit of Fig. 3-10B, for example, has a value of

$$v_o = i_c \times R_C \cong 10\ \mu A \times 5K \cong 50\ mV.$$

Thus the output voltage is greater than the input voltage. In fact, the ratio of the output voltage to the input voltage is the ratio of R_C to R_E, since the same current flows through both resistors. Mathematically, we can state the voltage gain (A_v) of circuit, *without* the emitter resistor bypassed, as

$$A_v = R_C/R_E.$$

This is a very useful equation; it shows that we can easily construct a circuit with any desired voltage gain we choose by just selecting resistors R_E and R_C. We need not consider the value of I_E or r_e for this circuit in calculating A_v. Of course, the circuit will only amplify as long as it is not driven into saturation, which means that for practical purposes, the available gain is usually fairly small.

EXAMPLE 3-5—In the circuit of Fig. 3-10B, change the collector resistor so that the gain of the circuit will be 8.

SOLUTION—
$$R_C = A_v \times R_E = 8 \times 1K = 8K$$

You can vary the gain by changing the emitter resistor, but remember that when you change the emitter resistor, you also change the input impedance.

It is also possible to control the gain of the amplifier by leaving just a *portion* of the emitter resistor unbypassed as in Fig. 3-11. In this circuit, the total dc resistance in the emitter is 2K, and since the dc voltage at the emitter is about 2 V, the dc emitter current is about 1 mA. But note that again the ac signal current which must flow through R_{E1} will develop a signal across R_{E1}, just as in the previous circuit. This same ac current will also flow through R_C, developing the ac output voltage across it. Thus the gain of the circuit is equal to the ratio of the collector resistance divided by the *unbypassed portion* of the emitter resistor, R_{E1}, or $A_v \cong 5K/500 = 10$. Again the input impedance seen looking into the base will be equal to $r'_{in} \cong \beta R_{E1}$.

Summarizing the characteristics of the amplifier of Fig. 3-10B, we see that the input impedance of this amplifier is much higher when the emitter resistor is left unbypassed, but the gain is reduced. However, the gain of the circuit depends only on the ratio of two resistors and is independent of any changes in transistor characteristics or changes in I_E.

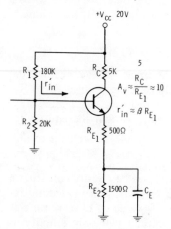

Fig. 3-11. Bypassing part of R_E increases ac gain.

The input impedance of the circuit is higher when R_E is not bypassed because when signal current flows, an ac signal is developed across the emitter resistor which bucks the applied signal. It is in phase with the applied signal and is applied to the emitter, reducing the drop across the base emitter junction. This causes the ac input current to be reduced. The scheme of using part of the output signal to increase the input impedance, or decrease the gain, or both, is called *negative feedback*.

Negative feedback can be applied over more than one stage to obtain excellent results. In the circuit of Fig. 3-12, we see a two-stage amplifier. As shown, amplifier stage Q1 has a gain of about 5 (if the loading cause by Q2 is negligible) and stage Q2 is shown to have a gain of 100. Note that the emitter resistor of Q2 is bypassed, so the gain of Q2 is determined by the emitter current. The overall circuit gain for both stages is about 500. Note the polarity of the waveforms at the various points in the circuit. The signal at the collector of Q1 is 180° out of phase with the input. Likewise, the signal at the collector of Q2 is 180° out of phase with the input of Q2, making it *in phase* with the original signal.

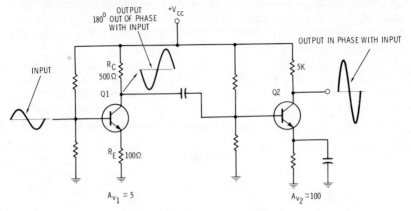

Fig. 3-12. Two-stage amplifier.

Fig. 3-13A is the same as the circuit of Fig. 3-12 except for an additional feedback component, R_F. Resistors R_F and R_E form an ac voltage divider across output of stage Q2 (Fig. 3-13B). We have assumed throughout that all capacitors have negligible reactance to the frequency we are working with. Notice that if the output signal appears from the collector of Q2 to ground, a fraction of the output voltage will appear across resistor R_E. The ac voltage across R_E will be

$$v_E = v_o \times \frac{R_E}{R_F + R_E}$$

We also know that, since the emitter resistor of Q1 is unbypassed, the signal appearing across emitter resistor R_E will be approximately equal to the input voltage, or $v_E \cong v_{in} \cong v_o \times R_E/(R_F + R_E)$. With v_{in} expressed in these terms, we can see that the gain of the entire circuit will be

$$A_v \cong \frac{v_o}{v_{in}} \cong \frac{R_F + R_E}{R_E}$$

Thus if R_F is much greater than R_E, a further simplification gives $A_v \cong R_F/R_E$. In order for these gain equations to be valid, the gain of the circuit without feedback, called the *open loop gain,* must be much greater than the gain with feedback, called the *closed loop gain.*

EXAMPLE 3-6—In the circuit of Fig. 3-13, if R_F is 2.5K and R_E is 100 ohms, what is the overall gain of the circuit?

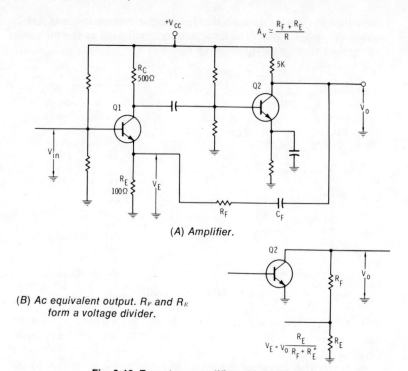

(A) Amplifier.

(B) Ac equivalent output. R_F and R_E form a voltage divider.

Fig. 3-13. Two-stage amplifier with feedback.

SOLUTION—

$$A_v = 2.5K/100 = 25$$

In this example, the open loop gain was given as about 500, making our calculated closed loop gain quite accurate.

This circuit is used extensively when higher values of gain are desired than can be obtained with a single stage, and where it is desired to make the gain stable—that is, independent of transistor characteristics or the values of current flowing in the transistors. Incidentally, negative feedback also gives the circuit a high input impedance.

Let's consider one more point before leaving this subject of cascaded amplifiers with feedback. In the circuit of Fig. 3-13, we can leave only a portion of the emitter resistor unbypassed in the first stage, and the gain will be determined by the ratio of feedback resistor R_F to the unbypassed portion of the emitter resistor. The *ac* equivalent circuit will still be the same as in Fig. 3-13B.

EXAMPLE 3-7—Using the configuration of Fig. 3-14, design a cascaded amplifier according to the thumb-rule design technique, described in Chapter 2, to set the dc collector current for each transistor at 1 mA. Choose R_4 so that the circuit will have an overall gain of 50. List all numbered resistance values.

SOLUTION—For $I_C = 1$ mA, and with a value of about 2 Vdc at each emitter, we will need a dc resistance in the emitter of each transistor of 2K. So, $R_9 = 2K$ and $R_4 + R_5 = 2K$. Next, we choose $R_2 = R_7 = 20K$ and $R_1 = R_6 = 180K$. The collector resistors should drop about half the remaining V_{CC} (or 9V) with 1 mA flowing, so let $R_3 = R_8 = 9K$ (or the closest standard value).

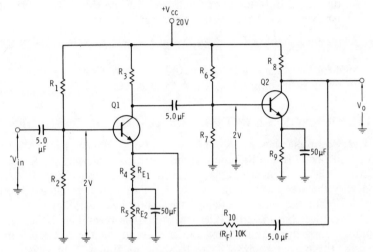

Fig. 3-14. Amplifier for Example 3-5.

Since we want the gain to be 50, the ratio of R_F to R_{E1} (R_4) should be 50. Thus $R_4 = R_F/50 = 10K/50 = 200$ ohms. Then to make $R_4 + R_5 = 2K$, we will choose $R_5 = 2K - 200 = 1.8K$.

Summarizing:

$R_1 = 180K$
$R_2 = 20K$
$R_3 = 9K$
$R_4 = 200\Omega$
$R_5 = 1.8K$
$R_6 = 180K$
$R_7 = 20K$
$R_8 = 9K$
$R_9 = 2K$
$R_{10} = 10K$

To build the circuit, you can use any general-purpose or audio transistors, if the circuit is intended for audio use. If you want more gain, make R_{10} larger. In fact, if you make R_{10} a pot, say 25K, you can use it as an excellent gain control.

63

Quiz

Design the following amplifiers for practice.

9. Using the circuit of Fig. 3-10B, change V_{CC} to 18 volts and select new values for all resistors so that $I_E = 0.5$ mA. The voltage across R_E is to be 1 volt, and the circuit gain is to be 12. For this circuit, make $R_2 = 10\ R_E$.

10. Using the circuit of Fig. 3-11, change the values of R_{E1} and R_{E2} so that the gain of the circuit will be 15. Keep the emitter current at 1 mA and leave the other resistors as they are.

11. Using the circuit of Fig. 3-14, assume V_{CC} is 14 volts. Design the amplifier to have an overall gain of 100 using the thumb-rule design for each stage. Let R_F be 10K as shown. Calculate all values of resistors, assuming I_C is 1 mA for each stage.

FREQUENCY RESPONSE OF RC-COUPLED STAGES

Every amplifier has some range or band of frequencies over which it operates best. In the RC-coupled amplifier, for example, if you connect a variable-frequency signal generator to the input (Fig. 3-15A) and vary the frequency from zero to some very high frequency, you will find that the output voltage varies somewhat as shown in Fig. 3-15B. Notice that at zero cycles, the output is zero. This is because coupling capacitors cannot pass direct current. As the frequency increases, the output begins to rise to some maximum value where it

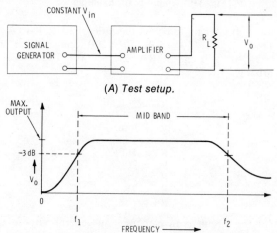

(A) Test setup.

(B) Typical response curve of RC-coupled amplifier.

Fig. 3-15. Determining frequency response.

levels off, then stays constant over a considerable range (called *mid-band*). Eventually, it begins to drop off again at some higher frequency. All of the gain equations we worked with earlier apply to the mid-band range, which is considered the useful operating range of the amplifier. In this section, we will look at some of the factors that make the output drop off at the low and high ends of the band.

The circuit of Fig. 3-16A will help us analyze what happens at low frequencies. This partial circuit represents the coupling network between an amplifier stage and a load, or between two amplifier stages. If we are looking at the coupling between two stages, resistor R_L represents the total input impedance to the next stage. In Fig. 3-16B, we see an equivalent circuit with a current source replacing the transistor. Finally, we can draw a Thevenin equivalent circuit of the current source and R_C (Fig. 3-16C), shown as a voltage source in series with R_C. Understanding the principle of the Thevenin equivalent circuit is not critical here, as long as you can accept that the circuits are equivalent.

As we can see in Fig. 3-16C, the coupling capacitor is simply in series with the two resistors. At mid-band and higher frequencies, the reactance of the coupling capacitor is negligible and no ac voltage will be dropped across it. However, as the generator frequency is lowered, the capacitive reactance increases by the formula

$$X_c = \frac{1}{2\pi f C_c}$$

where,

X_c is the capacitive reactance of C_c in ohms,
f is the signal frequency,
C_c is the capacitance in farads.

The lower the frequency, the greater the voltage drop across C_c. As more voltage is dropped across C_c, less is available to appear across R_L, so the output voltage decreases. As the frequency is lowered sufficiently, there will be some frequency at which the capacitive reactance is equal to the total series resistance. At this frequency, the power delivered to R_L will drop to half the maximum value, or we can say that the output is down 3 dB. This frequency is designated f_1 on the response curve of Fig. 3-15B. Mathematically, we can show that the lower 3-dB point occurs at a frequency $f_1 = 1/2\pi R_t C_c$, where $R_t = R_c + R_L$. If you want to make the amplifier perform better at lower frequencies, for example, you can either use a larger coupling capacitor or make the total series resistance larger, or both.

Emitter bypass capacitors also affect the low-frequency response of an amplifier stage. Rather than derive an equation to calculate exact values for the bypass capacitors, just remember that the capacitor is supposed to make the emitter grounded to ac; the larger the capacitor,

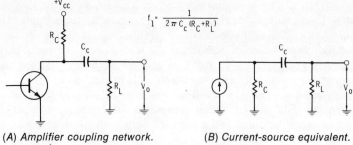

(A) Amplifier coupling network.

(B) Current-source equivalent.

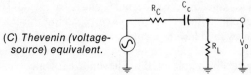

(C) Thevenin (voltage-source) equivalent.

Fig. 3-16. Circuits for modeling a low-frequency response.

the lower its reactance, and the better it grounds the emitter. Typically, if the emitter bypass capacitor has a reactance of about 100 ohms or less at the lowest frequency you intend to apply, it will not affect calculations.

Now consider the high-frequency end of the band. Fig. 3-17A shows the coupling network of an amplifier stage at high frequencies. The dotted capacitors are the collector-emitter capacitance (C_{ce}), the distributed wiring capacitance (C_w), and the input capacitance to the

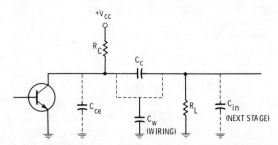

(A) Coupling network with distributed capacitance.

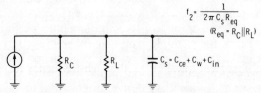

(B) High-frequency equivalent circuit.

Fig. 3-17. Circuits for modelling a high-frequency response.

following stage (C_{in}). All of these capacitances effectively appear in parallel across the load. In Fig. 3-17B, they are represented as a single shunt capacitance, C_s. Notice that coupling capacitor C_c does not appear in the equivalent circuit, since the reactance of this series-connected capacitor becomes negligible at higher frequencies. The reason that C_s is not bothersome at low frequencies is simply because it is usually a small value (say 100 pf or so), and such small capacitances have little effect on the signal except at high frequencies. However, because C_s is in parallel with the output, it will look more and more like a short circuit as the frequency is increased. When $X_{C_s} = R_{eq}$, the power delivered to R_{eq} drops to half the maximum value (3 dB drop). The mathematical formula for determining this upper 3 dB frequency is

$$f_2 = \frac{1}{2\pi R_{eq} C_s}$$

where,
f_2 is the upper 3 dB cutoff frequency,
R_{eq} is the effective parallel resistance of R_c and R_L,
C_s is the sum of capacitances C_{ce}, C_w, and C_{in}.

It is not extremely important to be able to calculate the exact value of f_2, but it is important to remember what quantities determine f_2. For example, if you want to increase f_2 and hence make the amplifier operate better at high frequencies:

1. Keep leads short to minimize wiring.
2. Use lower values of R_C or R_L if possible.
3. As a last resort, you may have to use transistors with lower input capacitance and lower output capacitance (rf type). Usually, the input capacitance of the following stage has the greatest effect on high-frequency response.

Quiz

Try the following quiz to see how well you understand frequency response. State *better, worse,* or *the same* for the following situations. Assume that you are working with an amplifier like that of Fig. 3-4.

12. If the value of a coupling capacitor is increased to twice the original value, the low-frequency response of the amplifier would be _____.

13. With the change of question 12, the high-frequency response would be _____.

14. If R_3 is reduced by 50%, the low-frequency response would be _____.

15. If R_3 is reduced by 50%, although the gain at mid-band would be less, the high-frequency response would be _____.

16. If Q2 in the circuit is replaced with a transistor having much less input capacitance, the low-frequency response would be _____.

17. If Q2 in the circuit is again replaced with a transistor having much less input capacitance, the high-frequency response would be _____.

18. If a 0.01-μF capacitor is connected from the collector of Q1 to ground, the low-frequency response would be _____.

19. With the change described in question 18, the high-frequency response would be _____.

GAIN CONTROLS

It is usually desirable to be able to increase or decrease the amplitude of a signal through an amplifier with some form of gain control or volume control. A gain control varies the gain of a circuit, while a volume control simply changes the signal level through an amplifier. For example, in Fig. 3-18A, emitter resistor R_E is a potentiometer. Since the gain of the stage is approximately R_C/R_E, varying R_E varies the gain. This circuit has a disadvantage, however, in that varying R_E also varies the dc collector current and thus changes the operating point so that it might cause distortion. An improvement over this circuit is shown in Fig. 3-18B. Notice that emitter resistor R_E is always a constant value as seen by direct current; however, by varying the position of the slider, more or less of the emitter resistor is bypassed through capacitor C_E. The ac gain of the stage is then the ratio of

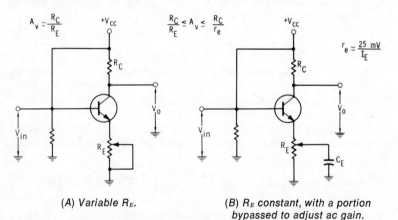

(A) Variable R_E.

(B) R_E constant, with a portion bypassed to adjust ac gain.

Fig. 3-18. Gain controls in the emitter circuit.

collector resistance (R_C) to the unbypassed portion of the emitter resistance, which is variable. The gain can be varied from a minimum value of R_C/R_E with none of R_E bypassed, to a maximum of R_C/r_e if all of R_E is bypassed ($r_e \cong 25$ mV/I_E).

A good gain control for use over more than one stage was discussed in the section on feedback and would be constructed by making R_F variable in Fig. 3-14.

Another way of varying the overall signal strength through an amplifier is to use a pot as a voltage divider, feeding only a portion of the signal to the following stage. In Fig. 3-19A, the collector resistor is replaced with a pot. By varying the pot slider, more or less of the ac signal developed across R_C will be fed to the following stage. This circuit, although simple and often satisfactory, has the disadvantage of varying the frequency response of the amplifier. You can see that this would be true by looking at the equivalent circuit of Figs. 3-16C and 3-17B again. Note that changing R_C in these circuits changes the frequency response.

A better solution, if a more consistent frequency response is desired, is that of Fig. 3-19B. Varying the pot slider applies more or

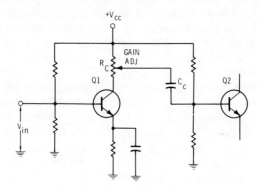

(A) *Volume control in the collector resistor.*

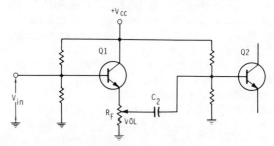

(B) *Volume control in the emitter circuit.*

Fig. 3-19. Volume control circuits.

less of the signal present at the emitter or the emitter-follower circuit to the following stage. If resistor R_E is reasonably small (a few hundred ohms), the overall frequency response will remain essentially unchanged at various gain settings.

MISCELLANEOUS DESIGN POINTERS

By now you have a pretty good idea of how various RC-coupled amplifier circuits work and what factors affect their operation. But you may still have some doubt as to when each circuit should be used, or how to select various other components for the amplifier. A few additional considerations will help you decide. Remember, we are talking only about small-signal (voltage) amplifiers here; power amplifiers will be discussed in a later chapter.

One of the first things to decide is the power supply voltage. This is not really a critical problem for small-signal amplifiers—use whatever is available. You can probably get higher gain without distortion by using a higher voltage, but for small signals this is usually of no great concern. A 9-volt transistor radio battery will work well for most small-signal amplifiers, as will a 12-volt source, if you have one. You can design the circuits discussed in this chapter to operate with almost any low-voltage supply of about 30 volts or less. Remember, if you use a very low voltage (say 5 V or less) you should take into account the few tenths of a volt drop across the base-emitter diodes when biasing the circuit.

Before deciding which circuit to use, you must decide what you want from the circuit. That is, do you want very high overall gain, maybe several thousand, or is a gain of 20 or 30 enough? Do you want maximum gain from the minimum number of stages for economy and small size, or do you need the gain to be stable and predictable, even if it costs more? The final design you select generally involves a compromise between several competing factors.

In general, to get maximum gain using the minimum number of parts, use the circuits with bypassed emitter resistors. A single stage with bypassed emitter resistor can produce a gain of 100 to 200 or so. By cascading stages as in Fig. 3-1, you can get an overall gain of several thousand. You can cascade up to perhaps three such stages with good results. More than three similar stages in cascade tend to make the amplifier unstable, with a tendency to oscillate.

If you want an amplifier with a stable, predictable gain, and one with low distortion, use circuits like Fig. 3-10B or Fig. 3-11. These will give gains of 20 or more, and if more gain is needed you can use a two-stage circuit (Fig. 3-14). These circuits also have a higher input impedance than ones using bypassed emitter resistors, which is usually an advantage.

Table 3-1 summarizes the characteristics of the amplifier circuits studied in this chapter.

Once you have determined the circuit configuration to use, you can start calculating resistor values. These calculated values are intended only as "ball park" figures. You can, of course, use the closest standard values for the final components. You may also want to modify the values slightly after testing the circuit. Resistors with a 5% or 10% tolerance will be satisfactory for the circuits described here.

Since the circuits are low-power, resistor wattage ratings can be small. Typical dc currents are generally in the neighborhood of 1 or 2 mA, so the power dissipated will be in milliwatts. You can use ⅛-watt or ¼-watt resistors if you need the small size, but ½-watt resistors will probably be the least expensive and easiest to obtain.

As for coupling capacitors, 2- to 10-μF capacitors will work fine in most audio-frequency circuits, and even smaller values are satisfactory at higher frequencies. The capacitors will usually be electrolytic, so you must observe the correct polarity when connecting them into the circuit. (For example, the collector of an npn transistor will generally be more positive than the base of the following stage.) Voltage ratings for the capacitors should be at least equal to the power supply voltage.

The emitter bypass capacitors must usually have a larger capacitance than coupling capacitors, probably 10 to 100 μF for most audio circuits. However, the voltage ratings will not have to be as large as for the coupling capacitors, typically less than half the power supply voltage.

Lastly, choose the transistors for the circuit. For audio amplifiers, any general-purpose or audio transistor will work fine. Manufacturers usually list transistors in catalogs according to their intended use. Higher frequency transistors, such as rf or switching types, certainly can be used for audio work, but audio transistors will probably not work well at radio frequencies. Be sure that the voltage ratings of the transistors are at least equal to the power supply voltage.

When selecting a transistor, remember that the price will not necessarily provide any information as to how good the transistor is for your purpose. Prices are usually determined by the quantity that the manufacturer sells, so a fairly inexpensive unit may give performance equal to that of a more costly one in many circuits. Also, remember that you can use either npn or pnp types, depending on whether you want to use a positive or negative power supply. Either type will work equally well.

Step-by-Step Design Summary for RC-Coupled Amplifiers

1. Choose a power supply voltage.
2. Select a suitable circuit based on your requirements from Table

Table 3-1. Summary of Transistor Amplifier Characteristics

Circuit	Possible Gain	Gain Stability	Input Impedance	Main Feature
Fig. 3-3	Up to 200	Fair	Rel Low, up to Few K	Maximum Gain for Single Stage
Fig. 3-4	Up to a Few Thousand	Poor to Fair	Rel Low, up to Few K	Used Where High Gain Is Needed
Fig. 3-5	1	Good	High	Used for Impedance Matching
Fig. 3-9	1	Good	Very High	Used for Impedance Matching
Fig. 3-10	Up to About 10	Good	High	Used Where Good Gain Stability Is Needed
Fig. 3-11	Up to About 25	Good	High	Same as Fig. 3-10, But Higher Gain
Figs. 3-13, 3-14	Up to About 100	Good	High	Used Where Fairly High Gain, Good Stability, Low Distortion Needed.

Note: component values should be chosen to meet specific requirements (see text).

3-1. (The values of components may have to be changed as needed.)
3. Calculate resistor values, based on both the "thumb-rule" design of Chapter 2 and the ac considerations of this chapter.
4. Choose coupling capacitors of 2 to 10 μF with voltage ratings equal to V_{CC}. Choose emitter bypass capacitors of 10 to 100 μF with voltage ratings of about ½ V_{CC}.
5. Choose any audio or general-purpose npn or pnp transistors, with voltage ratings at least equal to V_{CC}.
6. Build the circuit and test it. Make appropriate changes, if needed, based on your knowledge of how the circuit works.

Quiz

Test your ability to choose the right circuit for the job by selecting a suitable figure number for each example below. Disregard the values of components here; just choose the most suitable circuit configuration. Choose from Figs. 3-3, 3-4, 3-5, 3-9, 3-11, and 3-14.

20. You need a voltage amplifier with a stable gain of 12. You would like the circuit to have a fairly high input impedance, but it should not take up much space.
21. You need a fairly high-gain amplifier, about 2000 or so. Stability and input impedance are not important, but again you would like to keep the circuit reasonably small.
22. You need an amplifier with a stable gain of 75. It should have a fairly high input impedance.
23. You have a high-impedance microphone that you would like to connect to an existing amplifier. The problem is that the amplifier has a low input impedance, and when the microphone is connected to it, distortion results. You decide to put an impedance-matching stage between the microphone and the amplifier.
24. You want to build a preamp, to follow a low-impedance microphone, that can be built right into the microphone housing. This will increase the signal amplitude and cut down on the 60-Hz noise problem you are having. You are not interested in the exact gain of the circuit—the higher the better. But you do want to make the circuit as small as possible (using one stage).

SUMMARY

The input impedance to a stage affects the gain of the previous stage.

The input impedance of a stage varies inversely as the emitter current if R_E is bypassed.

The input impedance of a stage varies directly with β.

The gain of a stage varies directly with the dc emitter current if the emitter resistor is bypassed.

Emitter followers have unity voltage gain and a high input impedance. They are used to make low values of resistance look like higher values.

Leaving all or part of the emitter resistor unbypassed increases the input impedance of the stage.

The gain of a stage in which all or part of the emitter resistor is unbypassed depends only on the ratio of the collector load resistance to the unbypassed part of R_E; the gain is independent of I_E.

To improve the low-frequency response of an RC-coupled amplifier, increase the size of the coupling capacitor or increase the impedance of the circuit, or do both. (See Fig. 3-16.)

To improve the high-frequency response of an RC-coupled amplifier, decrease the shunt capacitance (C_{ce}, C_w, C_{in}), decrease the circuit impedance, or do both. (See Fig. 3-17.)

4

Field Effect Transistors

So far we have studied the bipolar, or junction, transistor. You will remember that the bipolar transistor is basically a current-controlled device. That is, output current I_C is controlled by input current I_B. Since some input current must flow, the input impedance seen by the driving source is fairly low, on the order of a few thousand ohms. We saw how to increase the input impedance of a stage by using an emitter follower; at best, however, the input impedance achieved might be on the order of a few hundred thousand ohms. Sometimes you have need of a device having an extremely high input impedance, say several megohms or more. This type of device is useful as an input stage for a voltmeter or scope, where you want minimal loading of the circuit being measured. One solid-state device having extremely high input impedance is the *Field Effect Transistor* (FET). Typical FETs have input impedances of hundreds of megohms or more and are therefore considered voltage-controlled devices. That is, *the output* **current** *of the FET is controlled by the input* **voltage,** since the input draws practically no current from the driving source.

There are two different types of FETs: the *Junction* FET, or JFET, and the *Metal-Oxide-Semiconductor* FET (MOSFET), also called the *Insulated-Gate* FET (IGFET). Each has its own advantage.

JFET CHARACTERISTICS

To make a Junction Field Effect Transistor we could start by making a bar of n-type semiconductor material. The fact that it is semiconductor material means that the bar acts somewhat like a resistor; if we connected a battery across the ends of the bar, some current would flow. The semiconductor bar is shown in Fig. 4-1A. One end of the bar, where electrons enter, is called the *source*. The other end, where they leave, is called the *drain*. Power supply V_{DD} determines the current flow through the bar. For example, if the resistance of the bar measured from S to D is 1K, and if V_{DD} is 10 volts, then the drain current (I_D) is 10 mA. Now let's look at how this resistor is changed into a JFET.

Fig. 4-1B illustrates the next step in the manufacturing process. A pellet of indium metal is placed on top of the bar of n-type semiconductor, then the bar is placed in an oven and heated. The pellet melts and soaks into the bar. The exact chemistry of what goes on is beyond the scope of this book, but the end result is that the region where the indium soaks in becomes p-type semiconductor (Fig. 4-1C). Since we now have p-type and n-type material next to each other in the same bar, we have a p-n (diode) junction. This is where the *Junction* FET gets its name. Finally, we connect a lead to the p-type material and call it the *gate*. It is the gate that will be used as the controlling element.

As was stated in an earlier chapter, the current carriers in n-type semiconductor are negative charges (electrons), and in p-type material the current carriers are positive charges (holes). Whenever a pn junction is formed, there is a region of transition between p and n materials where *no* current carriers exist. This is called the depletion region. Since no carriers exist, the depletion region is essentially a perfect insulator. It is this depletion region that makes the JFET possible.

In Fig. 4-2A we see the JFET connected to a power supply. Again electrons enter the source and leave through the drain. But this time, since the electrons cannot pass through a perfect insulator (the depletion region), they must all pass through the *channel* that exists be-

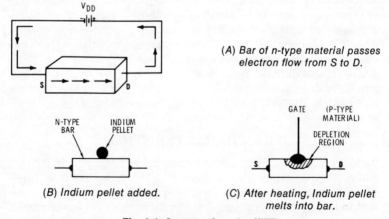

(A) *Bar of n-type material passes electron flow from S to D.*

(B) *Indium pellet added.*

(C) *After heating, Indium pellet melts into bar.*

Fig. 4-1. Construction of a JFET.

tween the depletion region and the side of the bar. So, effectively, what we have done by forming the depletion region is decrease the effective cross-sectional area of the bar through which current can flow. Reducing the cross section of a conductor or resistor increases the resistance. If we now connect the bar to the same source as in Fig. 4-1A, we find that the drain current is less than before. So we see that by changing the width of the channel, we can change the amount of current flow. Now here is how the JFET really becomes useful. If we connect a reverse biasing power supply between gate and source, as shown in Fig. 4-2B, we find that varying the power supply also varies the width of the depletion region. That is, the higher we make V_{GS}, the larger the depletion region becomes, and hence the narrower the channel. The end result is that by varying the gate-source voltage, V_{GS}, we cause the drain current to vary.

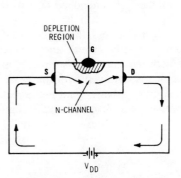 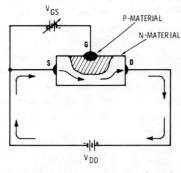

(A) Electrons must pass through the channel.

(B) Reverse bias across p-n junction spreads depletion region, reduces channel width.

Fig. 4-2. Current flow in a JFET.

Another important point to keep in mind is that the pn junction is reverse-biased; hence, no current flows in the gate lead. In practical circuits, the input signal voltage is usually applied between the gate and source. Since virtually no input current flows, the circuit has an extremely high input impedance.

When the drain current is made to flow through some load resistor, any variations in input voltage cause corresponding changes in output current; thus the JFET is basically a voltage-controlled device.

Schematic symbols and typical case diagrams for the JFET are shown in Fig. 4-3. Just as bipolar transistors are made in both npn and pnp types, there are two types of JFETs: n-channel and p-channel. The operating characteristics for either are essentially the same, except that in the p-channel version, the bar is made of p-type material and the gate is made of n-type material. All power supply polarities are

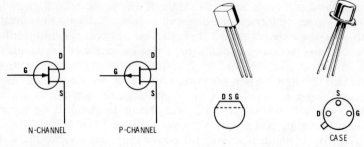

(A) Schematic symbols.   (B) Common packages for either type.

Fig. 4-3. Symbols and typical packages for JFETs.

reversed when using the p-channel device. Just as with the bipolar transistor, you can get a rough check of the type and condition of a FET using only an ohmmeter (Fig. 4-4).

An ohmmeter connected in either polarity across the source and drain leads measures the same resistance, usually 100 to about 10K ohms. When the meter is connected between gate and source, you will measure diode action. A low resistance is obtained with the negative lead of the meter connected to the source (assuming an n-channel device). When the connections are reversed, the result is a very high resistance, practically infinite.

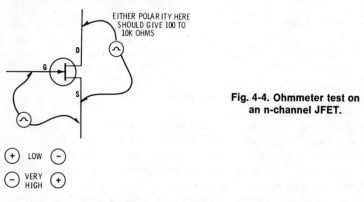

Fig. 4-4. Ohmmeter test on an n-channel JFET.

THE JFET AMPLIFIER

Just like the bipolar transistor, the JFET must be biased properly in order to operate. Fig. 4-5A shows an n-channel JFET biased with a reverse voltage between the gate and source, and a positive potential at the drain with respect to the source. These are the normal connections. The exact values of V_{GS} and I_D will depend on the specific device, but these need not concern us here. Suppose we find that the drain current is 4 mA, with $V_{GS} = -2$ volts. If this is the normal operating region for the JFET we would find that changing V_{DD} up or down a few volts has little effect on I_D. That is, the JFET acts somewhat like a constant-current source, similar to a bipolar transistor. However, the minimum voltage required between drain and source to make it function as a constant-current source is on the order of a few volts, compared to a few tenths of a volt for a bipolar transistor.

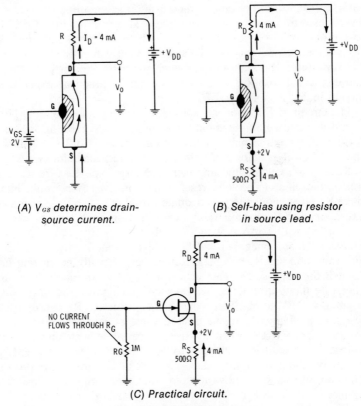

(A) V_{GS} determines drain-source current.

(B) Self-bias using resistor in source lead.

(C) Practical circuit.

Fig. 4-5. Biasing the JFET.

Now let's take a look at Fig. 4-5B. Notice that the gate is connected directly to ground, but this time the source is connected through a resistor to ground. Current flowing through the transistor from S to D must first flow through R_S. Now if R_S is 500 ohms as shown and it has 4 mA current through it, then the voltage drop across R_S will be 2 volts. Since the gate is at ground potential, the source is positive with respect to the gate; in other words, the gate is 2 volts negative with respect to the source. So the gate-source voltage is exactly the same as in Fig. 4-5A, without the need of an additional bias supply. The key to correct biasing is simply the proper choice of resistor R_S.

In Fig. 4-5C, we have the JFET in a circuit with the gate connected through resistor R_G to ground. Since no gate current flows in a JFET, there is no drop across R_G and the gate is still at ground potential. That is, in all three circuits of Fig. 4-5, the gate is 2 volts negative with respect to the source, so the biasing is essentially the same.

Remember from a previous chapter that you must not run a transistor into saturation or cutoff when using it as an amplifier; otherwise severe distortion will result. The same is also true for the JFET. It is best to operate the circuit about halfway between cutoff and saturation.

As stated before, you must choose the right value for R_S in order to bias the circuit. But without the specifications for the particular device, how can you choose R_S? Well, there is a simple, practical, experimental way of finding R_S for any JFET.

The circuit of Fig. 4-6 can be used to set up the bias on any JFET, even if you know nothing about the characteristics. Here again, we are dealing with small signals, so a drain current of 1 to 2 mA will work fine. Let $I_D = 1$ mA for convenience in making calculations.

Next, just as with any amplifier, the larger the load resistor, the greater the gain. (The load resistor here is R_D.) We would like to make R_D large enough to get adequate gain, but not so large as to cause saturation. Let R_D have a value that will drop about ⅓ to ½ of the V_{DD} supply with 1 mA flowing through it.

Assume that 1 megohm will be as high an input impedance as we will want, and use $R_G = 1$ meg. Remember, virtually no current flows through the gate, so the gate acts like an open circuit to the source. Therefore the total input resistance seen looking into the stage will be essentially equal to R_G. Somewhat higher values of R_G can be used if desired. Finally, set I_D to the desired value of 1 mA by connecting a pot in place of R_S. Adjust this resistance while watching the drain voltage (V_D) until is reaches ½ to ⅔ of the V_{DD} supply; a 25K pot should be about the right value. The circuit is now biased properly. You can, of course, remove the pot from the circuit, measure its resistance, and replace it with the closest, standard, fixed value. As a

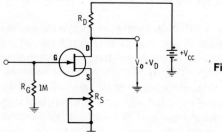

Fig. 4-6. Determining the correct JFET bias level.

last check, be sure that there are at least a few volts (say ⅓ of V_{DD}) left across the transistor from source to drain when V_D is set to the desired value. In other words, since half of the power supply voltage is dropped across R_D, be sure the other half is not dropped across R_S, or the transistor will be in saturation.

Summarizing the steps in biasing the circuit of Fig. 4-6:

1. Choose R_G equal to the desired input resistance.
2. Choose I_D (say 1 mA or so).
3. Choose R_D to drop ⅓ to ½ of V_{DD} with I_D flowing through it. (Typical values of V_{DD} are about 6-20 V or so.)
4. Adjust R_S until V_D reaches ½ to ⅔ of V_{DD}.
5. Replace pot R_S with the closest standard value of resistor.

EXAMPLE 4-1—Suppose you want to use a 12-volt power supply and choose 1-mA drain current for an amplifier stage. You decide to use a 5K drain resistor (R_D) and a 1 MΩ gate resistor (R_G). You put a pot in place of R_S and adjust it until the dc voltage at the drain is 7 Vdc with respect to ground. This will make the drop across R_D 5 volts; hence 1 mA will be flowing. You remove the pot, measure it to be 1.5K, and replace it with a 1.5K fixed resistor. The final circuit is shown in Fig. 4-7.

Note that the voltage across the JFET itself is $7 - 1.5 = 5.5$ V, so you don't have to worry about saturation.

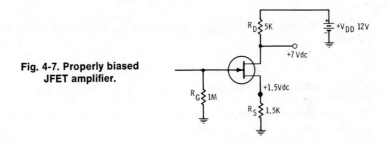

Fig. 4-7. Properly biased JFET amplifier.

A complete JFET amplifier is shown in Fig. 4-8. The input signal is applied between the gate and source, and the output signal is taken between the drain and source. (The source is grounded to ac by capacitor C_S, so the ac output is effectively taken between the drain and ground.)

In order to be able to calculate the gain of the stage, we must define a couple of quantities. Just as the dc drain current is controlled by the gate voltage, the ac drain current is controlled by the ac gate-source voltage. The ratio of ac drain current to ac gate-source voltage is called the *forward transconductance,* or *forward transadmittance.* That is,

$$y_{fs} = \frac{i_d}{V_{gs}}$$

where,
    y_{fs} is the forward transconductance in mhos,
    i_d is the drain current in amps,
    v_{gs} is the gate-source voltage.

The quantity y_{fs}, called g_m by some manufacturers, is the most important parameter of the FET for determining gain. (See the glossary at the end of this chapter for other characteristics.) Generally, the higher y_{fs}, the more gain you can get out of the amplifier. Typical values of y_{fs} are on the order of a few thousand micromhos.

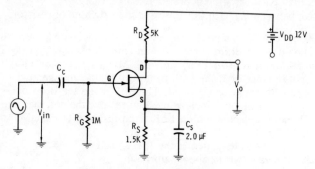

Fig. 4-8. Circuit for Example 4-2.

If you want to calculate the approximate voltage gain of an FET amplifier, use the equation

$$A_v \cong y_{fs} \times r_L \qquad \text{(Eq. 4-1)}$$

where r_L is the ac load resistance on the stage.

EXAMPLE 4-2—Using the circuit of Fig. 4-8, what is the voltage gain of the stage if $y_{fs} = 2500$ μmhos?

SOLUTION—In this case $r_L = 5K$, and from equation 4-1,
$$A_v = y_{fs} \times r_L = (2.5 \times 10^{-3}) \times (5 \times 10^3) = 12.5$$

Although FET amplifiers usually do not have a high voltage gain for a single stage, their big advantage is an extremely high input impedance. The input impedance of the circuit of Fig. 4-8 is essentially 1 MΩ. If you want an even higher input impedance, say 10 MΩ, just increase R_G to that value. The only limit to the size of R_G is that previously we assumed absolutely no current flow in the gate lead. In a practical FET, however, some current does flow, but the current is on the order of nanoamps, so small that it can be ignored unless R_G

is made very large (more than several tens of megohms). If R_G is too large, even a very small current flowing through it will upset the bias somewhat.

Quiz

To see how well you understand the JFET amplifier, try the following quiz. All questions refer to Fig. 4-9.

1. The polarity of the battery is (correct, incorrect).

2. If you want the amplifier to have an input impedance of about 2 MΩ, what value of R_G should you use?

3. Suppose the voltage from drain to ground measures 10 Vdc. You want to change this voltage to 7 Vdc. You would need to (increase, decrease) R_S.

4. If the generator feeds a signal of 20 mV ac to the gate, what is the approximate output signal voltage at the drain?

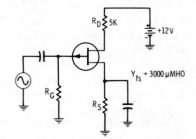

Fig. 4-9. Circuit for quiz problems.

MOSFETS

In order to get an even higher input impedance than is available with a JFET, the Metal Oxide Semiconductor FET (MOSFET) has no junction. As shown in Fig. 4-10A, the source and drain are made of heavily doped n-type material, and the channel between them is made of lightly doped n-type material. If a voltage is applied between S and D, some current will flow. By applying a negative voltage to the gate, electrons are driven away from the gate, effectively decreasing the channel width. The narrower the channel, the less current that will flow from S to D. This is similar to *depletion* in a JFET.

On the other hand, if a positive voltage is applied to the gate with respect to the source, more electrons will be drawn into the channel. The more carriers, the more current will flow between S and D.

Drawing additional carriers into the channel thus *enhances* the current flow through the device. The symbols for the n-channel and the p-channel MOSFET are shown in Fig. 4-10B.

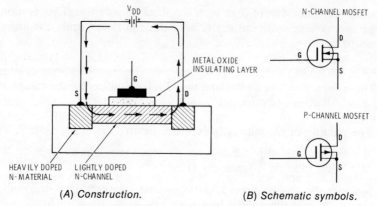

Fig. 4-10. Depletion-enhancement mode MOSFET.

To distinguish between the different methods, or *modes,* of operation, the *depletion* mode is also called mode A. FETs intended for the depletion mode only (JFETs) are called *depletion* transistors, or mode A transistors. The MOSFET with a lightly doped channel can be operated either in depletion or enhancement modes and is called a *depletion-enhancement,* or mode B, device.

There is still another method of making MOSFETs in which no channel is doped into the substrate. Fig. 4-11A shows that this device is similar to the mode B MOSFET, but the only way current can flow is by making the gate positive with respect to the source. This draws electrons up from the substrate (there are always a few free electrons at room temperature in the substrate) and thus *induces* a channel. This type of device is called an *induced channel* MOSFET, or *enhancement mode* (mode C) FET.

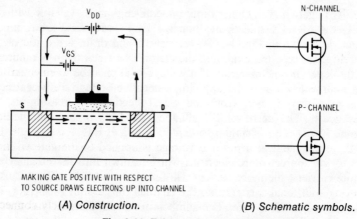

Fig. 4-11. Enhancement mode MOSFET.

Whether the MOSFET is built for mode B or mode C operation, virtually no current flows in the gate lead, since the gate lead is insulated from the channel by a thin layer of silicon dioxide, a nearly perfect insulator. The input resistance of the MOSFET is typically in the hundreds of megohms. However, since the layer of silicon dioxide is very thin, it can be punctured very easily with excessive voltage at the gate. (Typically, the maximum allowable voltage is on the order of 25 to 100 volts.) Even static electricity on your fingers is sufficient to ruin the device. For this reason, manufacturers supply MOSFETs with a shorting ring, or similar device, to prevent static electrical charges from accidentally ruining it. This shorting ring should be kept on the device until it is actually wired into the circuit. Also, the transistor should never be removed from the circuit with the power on, since accidental damage could result.

Since it is so easy to damage the MOSFET, it is not advisable to attempt to check the condition of one by switching ohmmeter leads back and forth. One way to get a rough idea of whether a MOSFET is in working condition or not is by using the setup shown in Fig. 4-12.

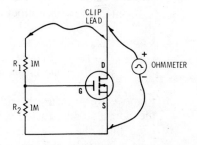

Fig. 4-12. MOSFET testing with an ohmmeter (see text).

If possible, connect resistor R_2 between G and S before removing the shorting ring. If this is not possible (some rings cannot be removed if the leads are soldered), try wrapping some metal foil around the case and the leads and leave it wrapped until the leads are soldered in place. Once R_2 is connected, attach the ohmmeter across the source and drain as shown, with the positive lead of the meter connected to the drain for an n-channel device (reverse this polarity for a p-channel device).

Before the upper end of R_1 is connected to the drain, you should measure from about 100 ohms to about 10K ohms for a mode B device, and infinity for a mode C device. Next, connect the upper end of R_1 to the drain with a clip lead. This will forward-bias either type somewhat, causing current to flow through the channel, and thus causing the ohmmeter to read a lower resistance (R_1 and R_2 in series). If the ohmmeter reading changes significantly as you alternately connect and remove R_1 from the drain, the transistor is probably all right.

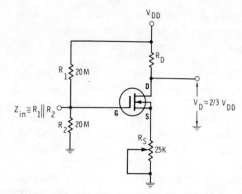

Fig. 4-13. Depletion-enhancement mode MOSFET amplifier.

To bias the MOSFET for operation as an amplifier, either circuit 4-13 or 4-14 will work for the type B (depletion-enhancement mode) device. Here is a short summary of how to bias the MOSFET.

The circuit of Fig. 4-13 can be used as follows:

1. Choose values of R_1 and R_2 so that the parallel combination will equal the desired input impedance. In the circuit shown, the input impedance would be 10 Meg.
2. Choose I_D (say 1 mA to 5 mA).
3. Choose R_D to drop about ⅓ of V_{DD} with I_D flowing through it.
4. Adjust pot R_S while measuring the dc voltage at the drain until $V_D \cong$ ⅔ (V_{DD}).
5. Remove R_S, measure it with an ohmmeter, and replace it with the closest standard fixed value.
6. As a final check, measure the voltage from S to D. If it is at least a few volts, the circuit is all right. If the voltage between S and D is too small, you will have to increase R_1 (make it larger than R_2) and readjust R_S again until V_D is the desired value.

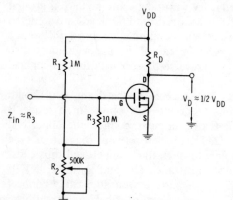

Fig. 4-14. Enhancement mode MOSFET amplifier.

The circuit of Fig. 4-14 can be used as follows:
1. Choose R_3 equal to the desired input impedance, r_{in}.
2. Choose I_D, again from about 1 mA to 5 mA.
3. Choose R_D to drop ½ of V_{DD} with I_D flowing through it.
4. Adjust R_2 while measuring V_D with a voltmeter until $V_D = ½ V_{DD}$.
5. Remove pot R_2, measure it with an ohmmeter, and replace it with the closest standard value.

(The input impedance of the circuit will be approximately equal to R_3, because no current flows into the gate lead.) The lower end of R_3 is returned to ground through the relatively small resistance of R_2.

Using the circuits of either Fig. 4-13 or Fig. 4-14, you can build small-signal amplifiers with gains determined by equation 4-1, just as with the JFET. If you use the circuit of Fig. 4-13, remember to bypass R_S with a capacitor as before.

FET APPLICATIONS

Since FETs have an extremely high input impedance, they are useful primarily as input devices. Because they do not have very high gains, it is usually best to use bipolar transistors to get high gain and use the FET merely as the input stage for an amplifier. In Fig. 4-15, for example, the JFET is used as a *source follower*. The source follower, like the emitter follower, has unity voltage gain. This circuit has an input impedance of about 10 MΩ, yet the FET can drive the relatively low input impedance of transistor Q2 without signal loss.

FETs can also be used to give at least some gain, while having a very high input impedance. The circuit of Fig. 4-16 makes a good preamp which can be used to connect a high impedance source, such as a condenser mike, to a low impedance transmission line. This circuit will amplify the input signal by at least ten or so, depending on

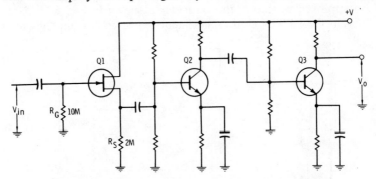

Fig. 4-15. JFET source follower gives high input impedance.

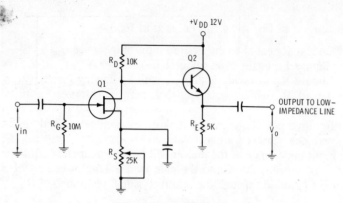

Fig. 4-16. FET preamp.

y_{fs} of the FET. The emitter-follower output of Q2 can drive a low-impedance load without decreasing the signal amplitude.

An important application of FETs is in transistor voltmeters. The tvm, like the earlier vtvm, has a high input impedance, but the tvm is smaller, more efficient, and can be operated with a low-voltage battery. A very simple FET tvm is shown in Fig. 4-17. To use the circuit, the input is grounded and R_7 is adjusted until the meter reads zero. Note that the circuit is similar to a balanced bridge; the FET and resistor R_3 form one side of the bridge, while R_6, R_7, and R_8 form the other side. By adjusting R_7, you are matching the voltage at the slider of the pot to the bias voltage between gate and source. Resistor R_2 is used to protect the FET in the event you accidentally apply the wrong polarity signal and forward-bias the gate-source diode. Note: if the meter will not zero with the values shown for the FET you are using, simply change R_8 up or down a little, and the circuit will balance. After the meter is zeroed, apply a known voltage of 1 volt to the input and adjust R_4 for full-scale deflection of the meter. The meter is now calibrated and ready for use.

An improved version of the voltmeter is shown in Fig. 4-18. Note that again the FET is used as the input to the circuit, thus giving a

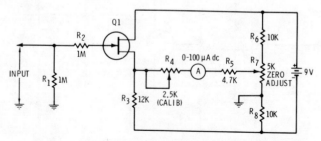

Fig. 4-17. Simple FET voltmeter.

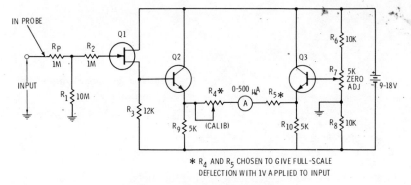

Fig. 4-18. Improved FET voltmeter.

high input impedance. The meter is actually driven by the emitters of Q2 and Q3, acting as emitter followers connected to the bridge. This circuit will be more accurate than that of Fig. 4-17 because there will be negligible loading on the bridge when it is unbalanced. With the circuit of Fig. 4-18, you can use a less sensitive meter, say 0-500 μA full scale, because the emitter followers will supply the deflection current. Resistors R_4 and R_5 are chosen to give full-scale deflection with 1 volt input, and calibration is accomplished by the same method used in the previous current.

To extend the range of the voltmeter and enable you to read higher voltages, R_1 can be replaced with a voltage divider as shown in Fig. 4-19. With the various resistors connected to different taps on a multiposition wafer switch, the voltmeter will have ranges of 1V, 3V, 10V, 30V and 100V full-scale deflection. The 1 MΩ resistor (R_P) is connected inside the probe to minimize capacitive loading of the cir-

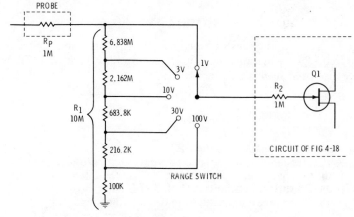

Fig. 4-19. Voltage divider extends voltmeter range.

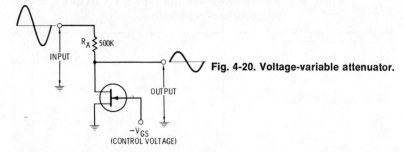

Fig. 4-20. Voltage-variable attenuator.

cuit under test. As before, after zeroing the meter, a 1-volt test voltage is applied to the probe tip and the calibration pot is adjusted to full-scale deflection. The values shown for the voltage divider is 1% tolerance for accuracy. If you do not need the accuracy, use the closest standard 5% values. All resistors are rated at ¼ watt.

One last circuit, which is very useful, makes use of the FET as a voltage-variable resistor. In Fig. 4-20, an input ac signal is applied to a voltage divider made up of resistor R_A and an FET in series. If the FET is reverse-biased, say to cutoff, it acts like an extremely high resistance; the output voltage is then about equal to the input voltage. But if the reverse bias on the gate is reduced, the FET resistance becomes smaller, so less and less of the input appears across the output. This circuit can be used as an agc control or voltage-variable attenuator. It is limited, however, to small ac input signals, because there is considerable distortion on larger signals.

The circuit of Fig. 4-20 can also be used as a chopper for either ac or dc input signals if the gate is driven with a negative-going square wave rather than a dc signal.

SUMMARY

FETs are used in circuits where an extremely high input impedance is desired. Two equations are applicable to either type of FET amplifier discussed in this chapter:

$$r_{in} \cong R_G$$
$$A_v \cong y_{fs} \times R_D$$

where,
r_{in} is the input resistance of the FET,
R_G is the resistance value connected from gate to ground,
A_v is the voltage gain,
y_{fs} is the forward transadmittance of the FET in mhos,
R_D is the resistance connected from drain to the power supply.

Thumb-rule design procedures are summarized in Table 4-1.

Table 4-1. Thumb-Rule Design for FET Circuits (Experimental Method)

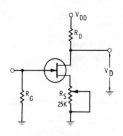

1. Choose R_G = desired r_{in}.
2. Choose I_D (1 mA or so).
3. Choose R_D to drop ⅓ to ½ V_{DD} with I_D flowing.
4. Adjust R_S until V_D reaches ½ to ⅔ V_{DD}.
5. Replace R_S with closest standard value.

Depletion Mode

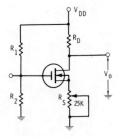

1. Choose $R_1 = R_2 = 2 \times$ desired r_{in}.
2. Choose I_D (1 mA or so).
3. Choose R_D to drop ⅓ of V_{DD} with I_D flowing.
4. Adjust R_S until V_D reaches ⅔ V_{DD}.
5. Replace R_S with closest standard value.

Depletion-Enhancement Mode

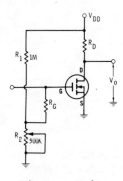

1. Choose R_G = desired r_{in}.
2. Choose I_D (1 mA or so).
3. Choose R_D to drop ½ V_{DD} with I_D flowing.
4. Adjust R_2 until V_D reaches ½ V_{DD}.
5. Replace R_2 with closest standard value.

Enhancement Mode

GLOSSARY OF FET TERMS

V_{DSS} — maximum voltage, drain to source.
BV_{GSS} — gate-to-source breakdown voltage when reverse-biased and $V_{DS} = 0$.
I_{GSS} — gate-to-source leakage current when $V_{DS} = 0$.
V_P — pinch-off voltage, gate-to-source voltage required to completely cut off drain current.
I_{DSS} — drain-to-source current with gate shorted to source.
y_{fs}, g_m — forward transadmittance (transconductance).

5

Differential and Operational Amplifiers

There are many cases in electronic systems when you will have to work with slowly changing signals or even small dc levels generated by thermocouples, servo amplifiers, photocells, or other transducers. If high gain is necessary, it is usually impractical just to cascade dc amplifier stages, because slight temperature changes or drifts will look like signals and also be amplified. One of the most popular high-gain dc amplifiers with good temperature stability is the *differential* or difference amplifier.

DIFFERENTIAL AMPLIFIERS

A simple differential amplifier is shown in Fig. 5-1. In this circuit two transistors have identical characteristics: $R_{B1} = R_{B2}$ and $R_{L1} = R_{L2}$. If R_{B1} is small, say about 20K or less, there is negligible voltage drop across it due to the small base current, so the voltage at the base of Q1 is approximately zero. Similarly, the voltage at the base of Q2 is about zero. The voltage at the emitters is also approximately zero (Q1 and Q2 being forward-biased); hence the current through R_3 is approximately equal to V_{EE}/R_3. Usually, R_3 is chosen to be large enough so that V_{EE} and R_3 together act like a constant-current source.

Assuming the transistors are identical, their emitter current will be about equal if $V_1 = V_2$. Therefore, the drops across R_{L1} and R_{L2} will also be equal, so that output voltage V_o will be about zero.

Suppose the temperature increases so that the leakage current of Q1 tends to increase. If the transistors are identical, and if they are mounted physically close together so that they are at identical tempera-

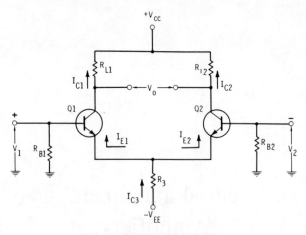

Fig. 5-1. Differential amplifier.

tures, both their leakage currents will tend to increase by the same amount. In fact, integrated circuit differential amplifiers have the transistors adjacent to one another on the same chip. This ensures that their characteristics will be very similar and that they will both operate at the same temperature. If both leakage currents increase slightly, the increased collector currents will cause equal voltage drops across the two collector resistors; the output voltage will still be zero. This is why the differential amplifier has good temperature stability. The output voltage does not depend on the exact value of collector current of any transistor, but on the *difference* between two collector currents. Therefore, temperature changes will not cause a significant difference in collector currents.

Next, suppose that V_1, the input voltage to Q1, goes slightly positive with respect to V_2. Since the base voltage of Q1 is more positive, it will conduct slightly harder. Both emitter currents join up to form I_3, and I_3 is fairly constant; so an increase in I_{E1} must cause a decrease in I_{E2}. Since Q1 conducts more, the drop across R_{L2} will be less. Therefore, output voltage V_o will be equal to the difference between the two collector voltages.

The input signal can be applied as shown in Fig. 5-2. The circuit is then said to have a differential input. Notice that neither end of the input signal is at ground. Sometimes you may want to amplify a signal generated by a device that must have one end grounded. In this case you will want to use a *single-ended* input, shown in Fig. 5-3. Notice that the base of Q2 is grounded, and since one end of the input generator is at ground, the input is still applied between the two bases.

You may also want the output signal to be referenced to ground. Fig. 5-4A shows a differential amplifier with a single-ended input and

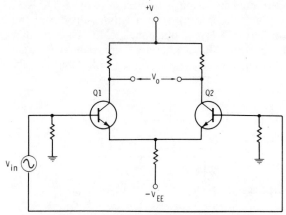

Fig. 5-2. Amplifier with differential input.

a single-ended output. When the input of this circuit goes positive, the output goes negative with respect to ground. This is like the normal 180° phase shift always associated with grounded-emitter amplifiers. But what if you feed the input signal to the base of Q2? When the input of such a circuit (Fig. 5-4B) goes positive, Q2 conducts more and causes the collector of Q2 to go negative, but the output taken at the collector of Q1 goes positive. In other words, the output and input of the circuit are *in phase*. Commercially available difference amplifiers are normally provided with two inputs: one is labeled the *inverting* input, and one the noninverting input. In Fig. 5-4, the input to Q1 is the inverting input, and the input to Q2 is the noninverting input.

It was mentioned that V_{EE} and R_3 form a type of constant-current source. Usually, a better constant-current source is desired and can

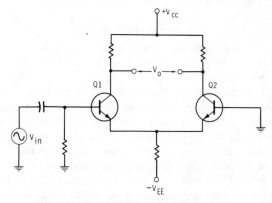

Fig. 5-3. Amplifier with single-ended input.

95

easily be constructed by using a transistor (Fig. 5-5). Resistors R_1 and R_2 (along with diode X1) form a voltage divider at the base of Q3. The drop across X1 is about equal to the base-emitter drop of Q3; the voltage across R_E is therefore approximately equal to the drop across R_2. Thus the voltage across R_E will determine the emitter

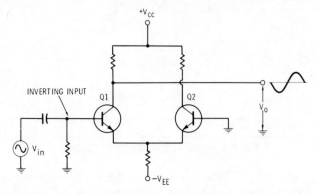

(*A*) With phase reversal.

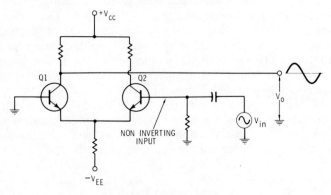

(*B*) Without phase reversal.

Fig. 5-4. Single-ended inputs and outputs.

current of Q3, which is, of course, the sum of I_{E1} and I_{E2}. Diode X1 is used as a temperature-compensating device. As the temperature changes, the base-emitter voltage drop of Q3 also changes, which would normally change the current through R_E. However, with X1 in the circuit, the drop across X1 changes by the same amount and in the same direction as V_{BE3}, keeping the voltage across R_E constant; hence the current through Q3 remains constant.

EXAMPLE 5-1—In the circuit of Fig. 5-5, let $R_1 = 5K$; $R_2 = 10K$; $R_E = 5K$; $R_{L1} = R_{L2} = 10K$; $V_{CC} = V_{EE} = 15$ volts; $R_{B1} = R_{B2} = 20K$.

Find I_{C1}, V_{C1} (measured from ground), and V_o. Assume Q1 and Q2 are identical.

SOLUTION—The current through R_1 and R_2 is approximately

$$I = \frac{V_{EE}}{R_1 + R_2} = 1 \text{ mA}$$

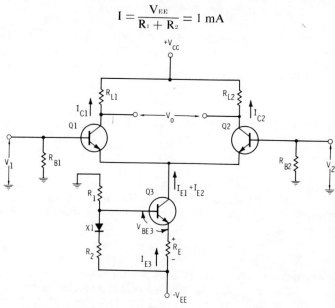

Fig. 5-5. Differential amplifier fed by constant-current source.

The voltage at the base of Q3 is equal to the 10-volt drop across R_2 plus the small drop across X1. The voltage across R_E must be approximately equal to the voltage across R_2, or 10 volts; thus

$$I_{E3} = \frac{10V}{R_E} = 2 \text{ mA}$$

Since the two transistors are identical, 1 mA flows through each, or $I_{C1} = 1$ mA $= I_{C2}$, and

$$V_{C1} = V_{CC} - I_{C1}R_{L1} = 15 - (1 \times 10) = 5V$$

V_o is equal to the difference between V_{C1} and V_{C2}, which is zero.

It can be shown that, for differential outputs, the voltage gain of the differential amplifier is

$$A_v = \frac{R_L}{r_e}$$

and for single-ended outputs,

$$A_v = \frac{R_L}{2r_e}$$

where,
$r_e = 25\text{ mV}/I_E$ and
I_E is the emitter current for either one of the transistors.

EXAMPLE 5-2—For the crcuit of Example 5-1, find the differential output voltage if $V_1 = 0.55$ volt, $V_2 = 0.53$ volt.

SOLUTION—
$$I_{C'} = I_E = 1 \text{ mA}$$
$$r_e = \frac{25 \text{ mV}}{1 \text{ mA}} = 25 \text{ }\Omega$$

Therefore
$$A_v = \frac{R_L}{r_e} = \frac{10K}{25} = 400$$

The differential input voltage is $V_1 - V_2 = 0.55 - 0.53 = 0.02$ V. The output voltage is then simply the input voltage times the voltage gain, or
$$V_o = V_{in} \times A_v = 0.02 \times 400 = 8 \text{ V.}$$

Feeding the emitters of Q1 and Q2 with a constant-current source has an additional advantage besides maintaining temperature stability.

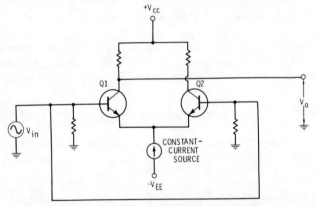

Fig. 5-6. Applying a common-mode signal to a differential amplifier.

Notice in Fig. 5-6 that the input signal generator is connected to both inputs in common. When the input signal goes more positive, *both* transistors tend to turn on harder at the same time by the same amount. However, the emitters are fed with a constant-current source, which means that the sum of the two currents I_{E1} and I_{E2} must be constant.

Even though both inputs are being driven more positive, neither transistor can turn on any harder, so the collector voltages of both transistors remain constant.

A signal applied to both transistors simultaneously is called a *common-mode* signal. This signal represents some unwanted signal, such as 60-Hz noise pickup, etc. Any unwanted signal will be applied to both transistors as though it were from a common generator and will not be amplified. The unwanted signal will usually be even smaller at the output than it is at the input. The ability of an amplifier to reject unwanted common-mode signals is called *common-mode rejection,* abbreviated cmr. This property is an important specification of differential amplifiers, especially in applications where the desired signal may be small compared to ambient noise.

The ratio of the differential-mode circuit gain to the common-mode circuit gain is called the *common-mode rejection ratio* (cmrr) and is expressed in decibels. For example, if an amplifier has a cmrr of 70 dB, and if equal common-mode and differential mode signals are applied to the input, the common-mode signal in the output will be 70 dB below the differential signal output.

The graph of Fig. 5-7 can be used to determine the voltage ratio, or gain, if the dB ratio is known. As can be seen from the graph, a ratio of 70 dB represents a voltage ratio of about 3200. So if equal differential and common-mode signals are applied, as above, the common-mode signal will be 3200 times smaller in amplitude than the differential signal at the output.

EXAMPLE 5-3—Motorola lists the differential gain of an MC1725 Differential Amplifier as 170, and the cmrr as 80 dB. If a differential signal of 2 mV is applied to the input along with an unwanted common-mode signal of 10 mV, what is the amplitude of each at the output?

SOLUTION—The differential output is simply the voltage gain times the input signal or

$$V_o = 2 \text{ mV} \times 170 = 340 \text{ mV}$$

The common mode gain is given by

$$\text{Common-mode gain} = \frac{\text{diff. gain}}{\text{cmrr}} = \frac{170}{80 \text{ dB}} = \frac{170}{10{,}000} = 0.017.$$

The amplitude of the common-mode signal is equal to the common-mode input signal times the common-mode gain, or

$$V'_o = 10 \text{ mV} \times 0.017 = 0.17 \text{ mV}$$

Notice that the differential signal is much larger at the output, while the common-mode signal has been greatly attenuated.

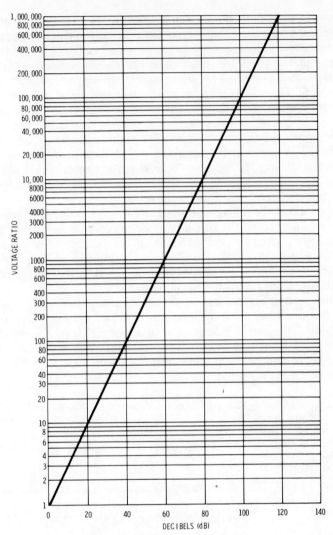

Fig. 5-7. Voltage ratio vs dB level.

OPERATIONAL AMPLIFIERS

If more gain is needed than can be obtained from a single differential amplifier, more stages can be cascaded and still maintain good temperature stability and a high common-mode rejection (see Fig. 5-8A). Notice that the two input terminals of this circuit are connected to Q1 and Q2. The outputs from these two transistors are connected to Q3 and Q4. Even more gain can be obtained by adding more stages, and the input impedance can be increased by adding emitter followers at the inputs. When connected in this manner, having high gain and high input impedance, the overall amplifier is called an *operational amplifier* or op amp. This name comes from earlier

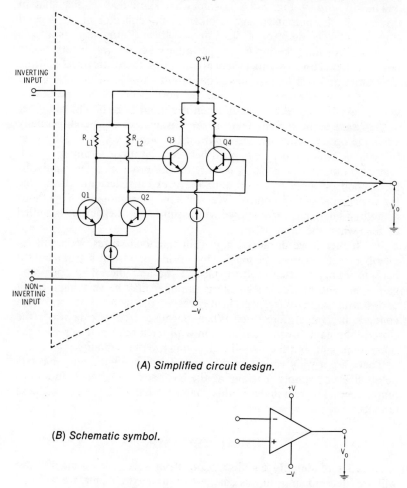

(A) Simplified circuit design.

(B) Schematic symbol.

Fig. 5-8. Cascading differential amplifiers to build on op amp.

work with these amplifiers, when they were used to perform mathematical operations. Now, due to their increased availability and reduced cost, they are used wherever stable, high-gain dc amplifiers are needed. The actual devices are usually integrated circuits and may differ from Fig. 5-8A by having more components than those shown, but the basic operation is still the same.

The symbol for the op amp is shown in Fig. 5-8B. By convention, the inverting input is labeled with a negative sign and the noninverting input with a positive sign. In addition, there are connections for the positive and negative power supplies.

The op amp of Fig. 5-8A is somewhat simplified. Notice that in this circuit, the output voltage is taken at the collector of the second stage (Q4). The dc level of the output voltage of any stage is always more positive than the input voltage; otherwise, the transistors would be saturated. However, in practical op amps there are level-shifting transistors and resistors to make the output capable of going positive or negative with respect to ground. In normal operation, the output voltage is adjusted to zero volts when the input is zero. The amplifier is then said to be *nulled*. One of the parameters of a commercially available op amp is called the *input offset voltage,* which is defined as the voltage which must be applied between the input terminals to obtain zero output voltage. It is usually on the order of a few millivolts for a good op amp. Some op amps have external terminals available for connecting a null control. We will talk more about the methods of nulling later, but for now, we will assume that the output is nulled to zero when the input is zero.

In Chapter 3 we discussed high-gain cascaded stages. You will remember that negative feedback (by means of resistors from output back to input) is often used to stabilize the amplifier. The open-loop gain (gain *without* feedback) of an amplifier may be very high, but it varies with slight differences in devices of the same type, and with changes in temperature, etc. When negative feedback is used, the closed-loop gain (overall circuit gain *with* feedback) is lower than the open-loop gain, but the amplifier has much better stability.

There are a wide variety of different circuit configurations that can be obtained by simply rearranging the feedback components in an op-amp circuit. The remainder of this chapter will explain several of these circuits.

IDEAL OP AMP

In order to simplify our discussion of various op-amp circuits, we will use ideal op amps in this chapter. In the next chapter, we will discuss practical limitations of op amps to see how they compare with ideal devices.

An ideal op amp is one having these characteristics:

1. Infinite open-loop gain.
2. Infinite input impedance looking into the terminals of the device.
3. Zero output impedance.
4. Zero dc output voltage when the input is grounded.

The open-loop gain of practical op amps is not really infinite, but it is extremely high, on the order of a hundred thousand or so, up to about one million for some devices. Next, the input impedance for commercially available op amps is not infinite, since the actual inputs are transistors and some bias currents must flow. But the input impedance is very high, possibly up to hundreds of megohms. Also, the output impedance can be less than 1 ohm in certain applications, so it can be considered negligible. Lastly, practical op amps can be nulled by means of external components, so that the output will be zero when the input is zero. For most practical purposes, we can consider commercially available op amps as being nearly ideal.

INVERTING AMPLIFIER

The circuit of Fig. 5-9 shows an inverting amplifier. In this, as in previous circuits, the ratio of the output voltage to the input voltage is equal to the voltage gain. That is, $A_v = V_o/V_i$. Solving for V_i we get $V_i = V_o/A_v$. Now if V_o has some finite value less than the power supply voltage (which is often about ± 15 V), and if the open-loop gain, A_v, is ideally infinite (or at least very large), we see that

$$V_i = \frac{V_o}{\infty} \cong 0$$

In other words, for any attainable value of output voltage under open-loop conditions, the input voltage required is so small we can say it is virtually equal to zero. We will use this approximation again later.

If an input voltage is applied to the left end of resistor R1, as shown in Fig. 5-9, it will cause a current to flow through it. Also, since V_i is approximately zero, terminal 1 on the amplifier is at about the same potential as terminal 2, namely at ground. We say that terminal 1 is at *virtual* ground, since it is not actually shorted to ground, but simply at ground potential.

The current through R_1 is then

$$I_1 \cong \frac{V_1}{R_1} \qquad \text{(Eq. 5-1)}$$

From this we see that the input impedance seen by the source is just resistance R_1, and we can build an amplifier with *any* input impedance we desire by simply choosing the value of R_1. That is, $Z_{in} = R_1$.

Now we can determine the output voltage. All of the current that flows through R_1 must flow through R_F, because the input impedance seen looking into terminal 1 of the amplifier itself is infinite. When I_1 flows through R_F, it causes a voltage drop across it equal to

$$V_o = I_1 R_F \qquad \text{(Eq. 5-2)}$$

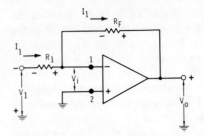

Fig. 5-9. Inverting amplifier.

This voltage *is* the output voltage, because the left end of R_F is at virtual ground and the right end is connected to the output.

Then dividing equation 5-1 by equation 5-2, we get the voltage gain of the circuit.

$$A_v = \frac{V_o}{V_1} \cong \frac{-I_1 R_F}{I_1 R_1} \cong \frac{-R_F}{R_1} \qquad \text{(Eq. 5-3)}$$

The negative sign indicates that the polarity of the output voltage is opposite from the input polarity.

EXAMPLE 5-4—Design an inverting amplifier with an input impedance of 1K and a voltage gain of 25.

SOLUTION—We simple choose a value for R_1 equal to the desired input impedance, namely 1K. Next, using equation 5-3, we calculate the value for R_F.

$$R_F = A_v R_1 = 25 \times 1K = 25K$$

The final circuit is shown in Fig. 5-10.

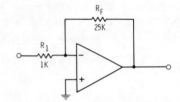

Fig. 5-10. Inverting amplifier with gain of 25.

NONINVERTING AMPLIFIER

Fig. 5-11 shows the diagram of a noninverting amplifier. Note that input voltage V_2 is applied between ground and terminal 2. Since the input signal is applied to the noninverting input, the output polarity will be the same as that of the input. Again we assume voltage $V_i \cong 0$, which means that a voltage equal to V_2 also appears across R_1. This causes current I_1 to flow through R_1 and, since no current flows into or out of the amplifier itself, the same current that flows through R_1

also flows through R_F. Finally, the left end of R_1 is grounded; thus output voltage V_o is the sum of the voltage drops across R_1 and R_F.

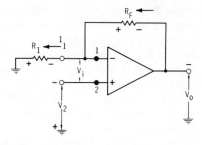

Fig. 5-11. Noninverting amplifier.

That is,

$$V_o = I_1 R_F + I_1 R_1 = I_1 (R_F + R_1)$$

If $V_i = 0$, the voltage across R_1 must be equal to V_2, or

$$V_2 = I_1 R_1$$

Solving for the voltage gain,

$$A_v = \frac{V_o}{V_2} = \frac{I_1(R_F + R_1)}{I_1 R_1} = \frac{R_F + R_1}{R_1} \qquad \text{(Eq. 5-4)}$$

The input impedance for the noninverting amplifier is much higher than for the inverting amplifier, since almost no current flows into input terminal 2. However, the input impedance of noninverting amplifiers changes with different devices of the same type, and with different gains. For now, just assume that the input impedance of a prac-

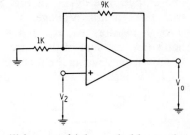

Fig. 5-12. Noninverting amplifier with gain of 10.

tical noninverting op-amp amplifier will be very high, probably on the order of several megohms.

EXAMPLE 5-5—Design a noninverting amplifier with a voltage gain of 10. Use a 1K resistor for R_1.

SOLUTION—Rearranging equation 5-4,

$$R_F = A_v R_1 - R_1 = 10 \times 1K - 1K = 9K.$$

The final circuit is shown in Fig. 5-12.

VOLTAGE FOLLOWER

The circuit of Fig. 5-13 is a *voltage follower*. The input signal is applied to terminal 2, producing an output voltage which is in phase with the input. The output is also tied directly to terminal 1, so it must be at the same potential as terminal 1. Since terminal 1 is always approximately at the same potential as terminal 2 ($V_i \cong 0$), the output voltage will be the same as the input voltage. The circuit therefore has unity voltage gain.

The voltage follower is used as an impedance-matching device, in the same manner as the emitter follower studied earlier. It has an extremely high input impedance, up to hundreds of megohms, and a low output impedance, often less than 1 ohm for practical amplifiers.

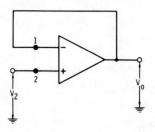

Fig. 5-13. Voltage follower.

SUMMER

Another useful circuit that can be built with an op amp is the *summer*, or *adder*, circuit. This circuit has two or more inputs and one output. The circuit of Fig. 5-14, for example, shows a summer with three inputs. Voltages V_1, V_2, and V_3 cause currents to flow through resistors R_1, R_2, and R_3, respectively. According to Kirchhoff's law, the sum of the currents entering a junction must be equal to the current leaving the junction. So the current which flows through R_F is the algebraic sum of the three input currents. Thus the output

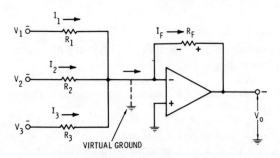

Fig. 5-14. Summer.

voltage is the result of all three input voltages added together algebraically. Mathematically,

$$V_o = -R_F \left(\frac{V_1}{R_1} + \frac{V_2}{R_2} + \frac{V_3}{R_3} \right)$$ (Eq. 5-5)

and under the special condition where $R_F = R_1 = R_2 = R_3$,

$$V_o = -(V_1 + V_2 + V_3)$$

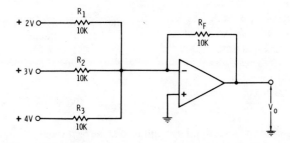

Fig. 5-15. Voltage summer for Example 5-6.

EXAMPLE 5-6—What is the output voltage of the circuit of Fig. 5-15?

SOLUTION—Since all resistors and voltage polarities are alike,

$$V_o = -(2 + 3 + 4) = -9 \text{ V}$$

INTEGRATOR

There is one more single-ended input circuit we want to consider. Fig. 5-16 shows an *integrator*. When voltage V_1 is applied as shown, current flows through R_1. Virtually no current flows into the op amp, so all of the current flows into the capacitor ($I_C = I_1$). Again, the inverting input is at virtual ground; thus $I_1 = V_1/R_1$. Now under the

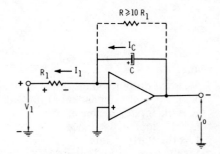

Fig. 5-16. Integrator.

special condition that V_1 remains constant, I_C remains constant. If a capacitor is charged with a constant current, a linearly changing voltage, or *linear ramp,* is developed across the capacitor.

The rate of change of output voltage with respect to time can be found by the expression:

$$\frac{\Delta V_o}{\Delta t} = \frac{-V_1}{R_1 C}$$

EXAMPLE 5-7—Suppose, in the circuit of Fig. 5-16, $R_1 = 100K$, $C = 0.1 \ \mu F$, and $V_1 = +3$ V. Find the rate of change of voltage across the capacitor.

SOLUTION—

$$\frac{\Delta V_o}{\Delta t} = \frac{-V_1}{R_1 C} = \frac{-3}{1 \times 10^5 \times 10^{-7}} = -300 \text{ V/sec}$$

If the output of the amplifier is initially at zero, the output waveform looks like Fig. 5-17. Note that the ramp is negative-going because the amplifier is inverting. If V_1 were a negative voltage with respect to ground, the ramp would be positive-going.

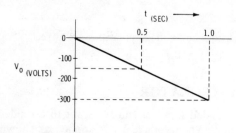

Fig. 5-17. Ramp voltage developed by integrator.

Developing a linear ramp voltage is not the only use for an integrator. In general, the output of the circuit will be the integral of the input, or

$$V_o = \frac{1}{R_1 C} \int_{t_o}^{t} V_1 dt$$

where,
V_o is the output voltage,
V_1 is the (time dependent) input voltage,
R is the input resistance,
C is the loop capacitance,
t_o is the starting time,
t is the total elapsed time.

The integrator is used largely in analog computers and also in waveform generators. Practical versions of the circuit usually include a resistor connected across capacitor C, as shown in Fig. 5-16. This resistor serves to keep op amp bias currents (discussed in the next chapter) from charging the capacitor.

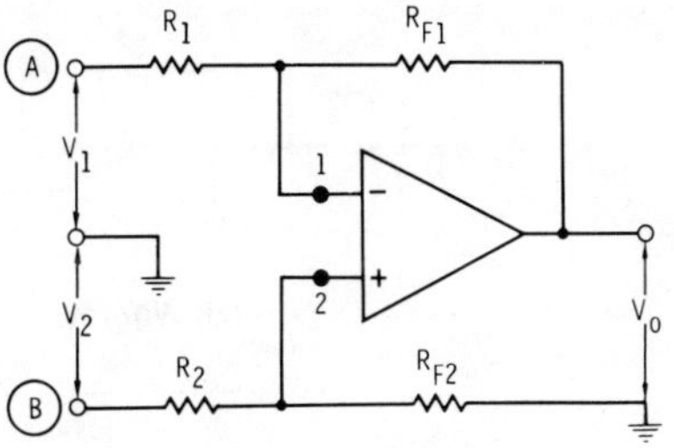

Fig. 5-18. Differential input amplifier.

DIFFERENTIAL-INPUT AMPLIFIER

So far we have used op amps only as single-ended input amplifiers. Many times you will have need for a differential-input amplifier. A differential-input amplifier is often used, for example, to minimize 60-Hz noise signals, such as in the input to an electrocardiograph. This instrument uses two electrodes fastened to different points on a human body and picks up minute electrical impulses generated each time the heart beats. These impulses are then amplified and fed to a speaker, scope, or strip chart recorder, for observation by a doctor. Unfortunately, besides picking up the desired impulses, a lot of 60-Hz noise is also picked up. This noise can be minimized by using a differential-input amplifier having a high common-mode rejection ratio. The differential input amplifier is shown in Fig. 5-18.

The differential-input amplifier is basically a combination of inverting and noninverting amplifiers. If $R_1 = R_2$ and $R_{F1} = R_{F2}$, the output of the amplifier can be shown to be

$$V_o = \frac{R_F}{R_1}(V_2 - V_1)$$

Remember, the output of the amplifier can go either positive or negative with respect to ground, so V_o may have either polarity depending on the polarities and magnitudes of V_2 and V_1.

EXAMPLE 5-8—In the circuit of Fig. 5-18, suppose $R_1 = R_2 = 1K$; $R_{F1} = R_{F2} = 100K$. If $V_2 - V_1 = 3$ mV, what is V_o?

SOLUTION—

$$V_o = \frac{100K}{1K} \times 3 \text{ mV} = 300 \text{ mV}$$

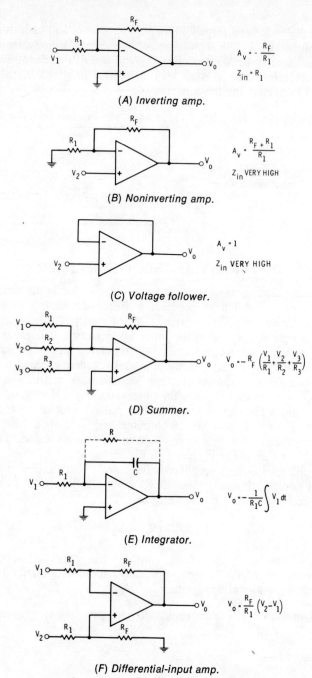

Fig. 5-19. Summary of basic op-amp circuits.

When actually using the differential-input amplifier, no ground connection is needed to the signal source if the two leads A and B are physically very close together. However, in electrocardiography, it is often desirable to use a third lead connected from the instrument ground to the patient's body. This ground lead is moved to various points on the body until the 60 Hz picked up at each lead is identical, thus nulling the 60-Hz output. Once noise has been nulled, the amplifier can pick up faint heart beat signals.

Quiz

1. In the circuit of Example 5-1, let $V_1 = 0.50$ volts, $V_2 = 0.55$ volts. What is the calculated value of V_o? Why will the actual measured value be smaller than this?
2. For the voltage summer circuit of Fig. 5-15, calculate the output voltage if R_F is changed to 5K (all other conditions the same).
3. Suppose, in the integrator circuit of Fig. 5-16, $R_1 = 10K$, $C = 0.01\ \mu F$, and $V_1 = 3$ V. Find the rate of change of voltage across the capacitor.
4. In the circuit of Fig. 5-18, what will be the value of V_o if $R_1 = R_2 = 1K$, $R_{F1} = R_{F2} = 300K$, and $V_2 - V_1 = 3$ mV? What will happen if R_1 and R_2 are not exactly equal, or if R_{F1} and R_{F2} are not exactly equal?

SUMMARY

In this chapter, we studied various op-amp circuit configurations using ideal op amps. In the next chapter, we will see how it is sometimes necessary to modify these circuits a little to compensate for non-ideal characteristics. However, the basic circuits and equations will still apply. Fig. 5-19 summarizes the basic circuits covered in this chapter.

6

Using Integrated Circuit Op-Amps

In the last chapter we discussed several circuits using ideal op amps. We will now discuss some of the limitations of commercially available integrated circuit op amps. All of the basic circuits we studied can be built with available devices, and all equations of circuit gain, etc., will still apply. The primary difference between practical and ideal op amps is that a few components are required to make actual circuits behave more ideally.

OUTPUT OFFSET CAUSED BY INPUT OFFSET VOLTAGE

You will remember that the output voltage, V_o, of an operational amplifier should be zero if the input to it is zero. However, you will find that for a practical IC op amp, if both inputs are zero, the output will not be exactly zero. That is, if both inputs of an op amp are shorted to ground as in Fig. 6-1, the output voltage will probably be *offset* slightly from ground. In some cases the output may even be saturated (equal to the power supply voltage). This output offset voltage, V_{oo}, is caused by an *input offset voltage,* V_{io}, which is the result of slight differences in characteristics of the input transistors on the chip. Basically, the circuit acts as though there were a small input voltage applied to the noninverting input, as shown in Fig. 6-2. The input offset voltage is multiplied by the gain of the amplifier.

Manufacturers list the input offset voltages for various devices. The value is typically just a few millivolts.

EXAMPLE 6-1—If the input offset voltage of the op amp in Fig. 6-2 is 5 mV and the amplifier has an open-loop gain of 100,000, what will be the magnitude of the output offset voltage?

SOLUTION—
$$V_{oo} = A_v \times V_{io} = 100{,}000 \times 5 \text{ mV} = 500 \text{ V}$$

Of course, the output voltage would not go as high as this. Typically, power supply voltages are ±15 V, so when the output reaches either extreme of the supply voltage it simply remains there.

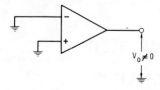

Fig. 6-1. Practical op amp has some output voltage when both inputs are grounded.

If the circuit uses feedback to reduce the gain, there will still be some value of output offset voltage. The output offset voltage will simply be equal to the input offset voltage times the closed-loop gain for the amplifier. That is,

$$V_{oo} = \frac{R_1 + R_F}{R_1} \times V_{io}$$

EXAMPLE 6-2—In the circuit of Fig. 6-3, find the output offset voltage if V_{io} is 5 mV.

SOLUTION—
$$V_{oo} = \frac{1K + 10K}{1K} \times 5 \text{ mV} = 55 \text{ mV}$$

In this case the output offset voltage is small enough to be considered negligible for many applications, but remember that the larger the closed-loop gain, the more bothersome the input offset voltage will be.

For some applications, especially in high-gain dc amplifiers, it is important that the output be exactly zero when the input is zero. The output can be adjusted to zero by means of a *null control* as shown in

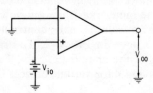

Fig. 6-2. Op amp offset can be represented by a small voltage applied to noninverting input.

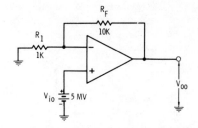

Fig. 6-3. Amplifier for Example 6-2.

Fig. 6-4. (The +V and −V sources are the normal power supply connections for the circuit.)

Many op amps come supplied with terminals for connecting a null pot, as shown in Fig. 6-5. Regardless of the value of V_{io} or A_v, the output can be adjusted to zero by moving the slider on the null control.

Quiz

For the following questions, refer to Fig. 6-6.

1. If $R_1 = 2K$, $R_F = 10K$, and $V_{io} = 4$ mV, what is V_{oo}? $V_{oo} = $ _____.

2. If $R_F = 200K$, $R_1 = 40K$, and $V_{io} = 4$ mV, what is V_{oo}? $V_{oo} = $ _____.

3. If $R_F = 200K$, $R_1 = 10K$, and $V_{io} = 4$ mV, what is V_{oo}? $V_{oo} = $ _____.

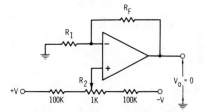

Fig. 6-4. Null control R_2 sets V_o to zero.

OUTPUT OFFSET CAUSED BY INPUT BIAS CURRENT

Besides input offset voltage, there is one other factor that can cause the output to be offset. Since the input stages to the op amp are transistors, bias currents must flow into the input leads. If an amplifier is connected as in Fig. 6-7, bias current I_{B1} flowing through R_1 causes a voltage to develop across R_1. This voltage acts as a differential signal and is amplified. The magnitude of the output voltage, V_{oo}, generated by I_B can be shown to be

115

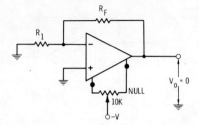

Fig. 6-5. Op amp with external terminals for null control.

$$V_{oo} \cong R_F \times I_B$$

The value of I_B is given by manufacturers and is usually in the order of a microamp or less. It may be a fraction of a nanoamp for op amps with FET input transistors.

EXAMPLE 6-3—If I_B in the circuit of Fig. 6-7 is 1μA and R_F = 200K, what is V_{oo}? If R_F is increased to 10 MΩ, what is V_{oo}?

SOLUTION—When R_F = 200K,

$$V_{oo} = 200K \times 1\ \mu A = 2V$$

When R_F = 10 MΩ,

$$V_{oo} = 10\ M \times 1\ \mu A = 10V$$

It is evident that larger values of R_F increase the effect of I_B.

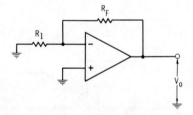

Fig. 6-6. Circuit for quiz problems.

Since the op amp is a differential amplifier, we can cancel the effect of I_B by applying a voltage to the noninverting input equal in magnitude to the voltage developed across R_1 by I_B (Fig. 6-8). Connecting

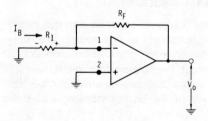

Fig. 6-7. Bias current at the inverting input.

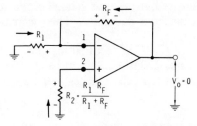

Fig. 6-8. Resistor R_2 minimizes the effect of input bias current.

resistor R_2 from the noninverting input to ground, the resulting current through R_2 will develop a voltage across R_2. Resistor R_2 is chosen so that the voltage developed across it is equal to the voltage developed across R_1. Usually resistor R_2 must be slightly smaller than R_1 because not all of I_B flows through R_1; part of it flows through R_F if the output is near ground. The correct value of R_2 is given by

$$R_2 = \frac{R_1 R_F}{R_1 + R_F}$$

Sometimes the input bias currents are not exactly equal; the difference between the two is known as the *input offset current*. So even if resistor R_2 is used, the output might not be *exactly* zero, but it is usually small enough for most applications. In the event you must have the output voltage exactly equal to zero, as in an analog computer, for example, the output can be nulled to zero with one of the null circuits discussed previously.

Quiz

4. Design an inverting amplifier similar to the one in Fig. 6-8, having an input impedance of 2K ohms, and a voltage gain of 25. Ans. $R_1 = $ _____, $R_F = $ _____, $R_2 = $ _____.
5. For large value of gain, say 20 or more, the value of R_2 can be approximately equal to R_1. (true, false)
6. It is *always* necessary to use some sort of variable null control when using an op amp. (true, false)

FREQUENCY RESPONSE

As you may recall from the chapter on cascaded transistor stages, the output of an amplifier begins to fall off at some high frequency. This decrease in gain at high frequencies is due primarily to shunting capacitance distributed throughout the amplifier. Fig. 6-9A shows a circuit representing the output of a high-gain amplifier. At low frequencies, output voltage V_o is equal to V_a because the capacitors are small and have negligible effect. However, as the frequency is increased, the output voltage begins to drop, because X_C of the shunting capacitors decreases. This decrease in output voltage is shown in Fig. 6-9B.

Besides the fact that the amplitude of the output drops, there is also a phase shift of V_o with respect to V_a. At some frequency the phase shift eventually reaches 180°.

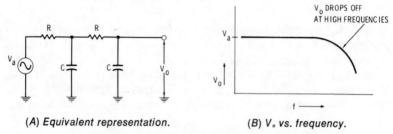

(A) *Equivalent representation.*   (B) *V_o vs. frequency.*

Fig. 6-9. Amplifier output at high frequencies.

Here is where the problem of oscillation arises. In an inverting amplifier like that of Fig. 6-10, the phase relationship between V_1 and V_o is normally 180°; but at some frequency there is an additional 180° phase shift at the output, caused by the effect of shunting capacitance. This means that the output is 360° out of phase with V_1, or actually *in phase* with V_1. This will cause the circuit to go into oscillation. (Further discussion of why circuits oscillate is given in Chapter 7.)

To prevent op amps from going into oscillation, *frequency compensating* components are connected between various points in the circuit. These compensating components decrease the gain of the op amp sufficiently at frequencies where a 360° phase shift occurs, so that the op amp cannot oscillate. It is not really necessary to understand how these components work, because manufacturers tell you exactly what values to use and where to connect them. The values of compensating

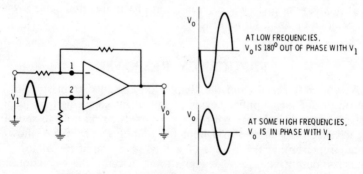

Fig. 6-10. Phase relationship of output to input.

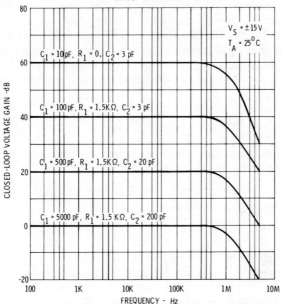

Fig. 6-11. Frequency response curves for the 709 op amp.

components are chosen depending on the desired gain and maximum frequency to be amplified.

Fig. 6-11 shows various frequency response curves for a 709 op amp. For example, if you need a gain of 40 dB, the curve shows that you should use values of $C_1 = 100$ pF, $R_1 = 1.5$K, and $C_2 = 3$ pF. The connections for the compensating components on the 709 are shown in Fig. 6-12.

Some op amps are internally compensated and do not need the external components. With internal compensation, the gain falls off at lower frequencies than with the externally compensated amp. Fig. 6-13 shows the frequency response of curve for the 741 op amp, which is internally compensated. Notice that the open-loop gain drops from its dc value of over 100,000 to unity at 1 MHz.

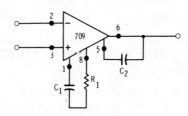

Fig. 6-12. Pin connections for frequency compensation of the 709 op amp.

For a gain of 100 (40 dB), the maximum frequency that the 741 op amp will work at is 10 kHz. Using the 709, you can see from Fig. 6-11 that with 40-dB gain, the output just begins to drop off at about 500 kHz. Comparing the two types then, we see that if you are going to work only with low frequencies, the internally compensated amplifier is easier to use, since there are fewer components to wire. But if you need good response at high frequency, use an externally compensated device.

Quiz

7. You want to build an amplifier with a gain of 1000 (or 60 dB). What is the maximum frequency response of the amplifier if you use a 741? $f_{max} = $ _____.
8. If you use a 709 instead of the 741 in problem 1, what is the maximum frequence response? $f_{max} = $ _____.

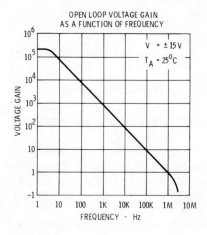

Fig. 6-13. Frequency response of the 741 op amp.

9. If you use the 709 as in problem 2, what values of compensating components would you use?
   $R_1 = $ _____, $C_1 = $ _____, $C_2 = $ _____.
10. You are building a voltage follower to operate up to 500 kHz. Which device, 709 or 741, would be a better choice? (Assume that the cost is about the same for either device.)

POINTERS ON WORKING WITH IC OP AMPS

Commercially available op amps come in a few different types of packages, the most popular being the TO-99, shown in Fig. 6-14A, and the dual in-line package (DIP) shown in Fig. 6-15A. The DIP shown is the 8-pin plastic case, but op amps are also available in other DIP configurations. The 709 and the 741 are currently the two most popular IC op amps. Also shown in Figs. 6-14 and 6-15 are the pin connections for these two devices. Various manufacturers may label the devices slightly differently; for example, Motorola labels one device MC 1709 and Fairchild calls it μA709, but the main part of the number is 709, and the devices are basically the same.

Choosing a Device

1. When working with either dc or low-frequency signals, use the 741. (An externally compensated device can be used, but it requires extra components.)
2. When building voltage followers, use the 741. (Again, fewer components are needed.)
3. When designing high-gain circuits for high-frequency applications, use the 709.

Choosing a Circuit

1. Inverting amp—
    (A) Use when inversion is desired.
    (B) Use when constant input impedance of low to medium specified value is required ($Z_{in} = R_1$).
2. Noninverting amp—
    (A) Use when no phase reversal is desired.
    (B) Use when high input Z is desired (Z is not constant).
3. Voltage Follower—Use for impedance matching when extremely high input Z and low output Z are required.
4. Summer—Use to add two or more inputs algebraically.
5. Differential Input—Use to eliminate some common-mode voltage or noise (60 Hz, etc.) from signal.
6. Integrator—Use to integrate, generate linear ramp, or produce triangular waves.

We will discuss several applications of op amps in the remaining chapters of this book.

SUMMARY

Practical integrated circuit op amps can perform almost like ideal devices. However, commercially available op amps have an output offset when the inputs are zero, which is a problem whenever the circuit has high gain.

Output offset can be minimized by putting a resistor in the noninverting lead. However, even with a resistor the output may still be offset somewhat, and this can be troublesome with large values of feedback resistance. The remaining output offset, caused by either input offset voltage or input bias current, can be reduced to zero using a null control. Nulling can be applied either at the input or preferably at the null terminals, if provided.

Practical amplifiers must be compensated to prevent them from oscillating. Some devices are internally compensated; others must be externally compensated.

When designing high-gain amplifiers for high-frequency work (1 kHz and up), use externally compensated devices. Otherwise, use internally compensated devices.

Design Quiz

11. Design an amplifier with a gain variable from 1 to 100, to operate at frequencies up to 400 Hz. The amplifier should have a constant input impedance of 500 ohms to match a 500-ohm trans-

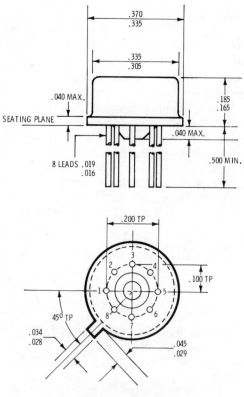

(A) Case dimensions (inches).

Fig. 6-14.

ducer. Choose either a 709 or 741. Choose an appropriate circuit. Draw the diagram, label the pin numbers, and show all part values.

12. Design an amplifier with a gain of 1000 to handle frequencies up to 100 kHz. The amplifier should have an input impedance of more than 100K ohms, but the input impedance need not be constant. Draw the complete circuit as in Problem 11.

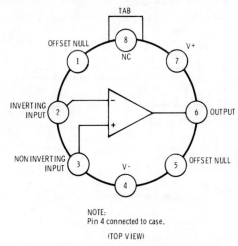

(B) 741 pin connections.

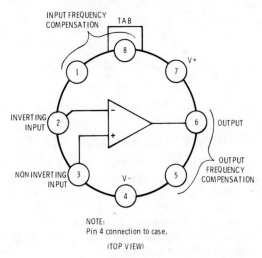

(C) 709 pin connections.

TO-99 package.

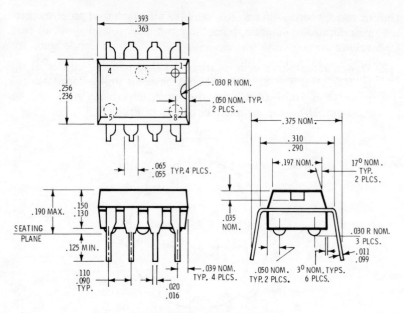

(A) Case dimensions (inches).

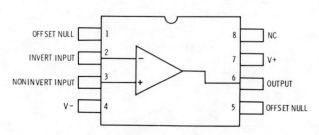

(B) 741 pin connections.

Fig. 6-15. DIP package.

GLOSSARY OF OP AMP TERMS

Input offset voltage—The voltage which must be applied between the input terminals to obtain zero output voltage.

Input offset current—The difference in the currents into the two input terminals with the output at zero volts.

Input bias current—The average of the two input currents.

Input resistance—The resistance seen looking into either input terminal with the other input grounded.

Input voltage range—Range of voltage on input which, if exceeded, could cause malfunction.

Output voltage swing—The peak output swing, referred to zero, that can be obtained without clipping.
Common-mode rejection ratio—Ratio of differential-mode gain to common-mode gain.
Slew rate—Maximum rate of change of output voltage with step input voltage.

7

Oscillators and Waveform Generators

In the past, sine-wave oscillators were usually built with LC-tuned circuits. More recently, an increasing number of audio frequency oscillators and waveform generators are being designed without inductors. The advantages of RC circuits over LC circuits are reduced size and, often, reduced cost, especially at very low frequencies where suitable inductors are large and expensive. In this chapter we will discuss a few types of RC oscillator circuits.

PHASE-SHIFT OSCILLATOR

Fig. 7-1A shows a simple block diagram of an oscillator. Basically, all that is needed is an amplifier and a feedback network. Fig. 7-1B shows the block diagram in a little different fashion to explain how the oscillator works. The generator feeds a signal, V_1, to the input of an amplifier (point A). The amplifier can be a simple inverting amplifier, such as the single-stage, common-emitter amplifier studied in Chapter 2. Notice that the output from the amplifier, V_2, is larger in amplitude and inverted (180° out of phase with V_1). The output from point B is then fed through a phase-shifting network which alters the phase of V_2 by 180°, making it *in phase* with V_1. The output (V_3) from this phase-shifting network appears at point C. If the signal at point C is of the same amplitude and phase as V_1, point C can be connected to point A and the circuit will continue to have an output, even if the generator is removed from point A. That is, the circuit will be oscillating.

Now we can look at the phase-shifting network. Fig. 7-2A shows an RC circuit with generator V_g applying a voltage to the input. At

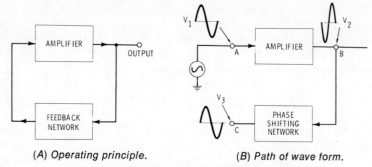

(A) Operating principle.   (B) Path of wave form.

Fig. 7-1. Block diagram of a simple oscillator.

frequencies where X_C is not negligible, the voltage appearing across the resistor will be reduced in amplitude and shifted in phase from the generator voltage (see Fig. 7-2B). If the generator frequency is varied, the phase angle between V_g and V_R will also vary. At some frequency, the relative phase angle will be 60°.

Next, consider the circuit of Fig. 7-2C. The generator V_g applies a voltage to the input of the network. At some frequency, the voltage V_{R1} will lead V_g by 60°. If $C_1 = C_2 = C_3$, and if $R_1 = R_2 = R_3$, the voltage V_{R2} will lead V_{R1} by 60° (leading V_g by 120°), and V_{R3} will lead V_{R2} by 60° (leading V_g by 180°). So, by using the three RC

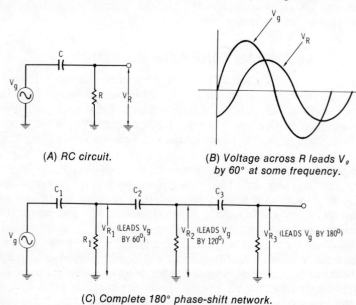

(A) RC circuit.   (B) Voltage across R leads V_g by 60° at some frequency.

(C) Complete 180° phase-shift network.

Fig. 7-2. RC phase-shift networks.

sections in series, output voltage V_{R3} will be 180° out of phase with the input voltage.

We can use the phase-shifting network of Fig. 7-2C along with an amplifier to build an oscillator. The circuit of Fig. 7-3 is a complete phase-shift oscillator. Notice that the bottom end of R_3 is not grounded in this case, but is connected instead to the input of the amplifier. The signal fed into the base of the transistor will be 180° out of phase with the collector voltage, since the current through R_3 is in phase with the voltage across R_3. Thus the current through R_3 will develop a signal voltage across the input to the transistor which is of the proper phase relationship to maintain oscillation. No signal is needed to start the circuit oscillating; if the gain is sufficient, it will begin oscillating as soon as power is applied to it.

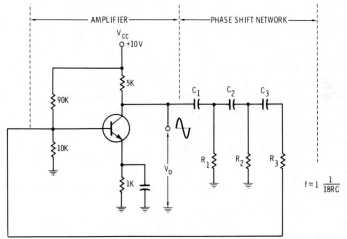

Fig. 7-3. Phase-shift oscillator.

A rough estimate of the oscillator frequency can be determined by the formula

$$f = \frac{1}{18\,RC}$$

Where $R = R_1 = R_2 = R_3$ (maximum value for each resistor should not be greater than 10K), and $C = C_1 = C_2 = C_3$.

By replacing R_2 with a pot, the oscillator frequency can be varied over a small range.

Some experimenting with this circuit may be necessary to achieve stable oscillation; that is, you may need to adjust the gain or biasing a little to make it oscillate. The circuit works well at audio frequencies, but it may be difficult to get it to oscillate at 100 kHz or above.

EXAMPLE 7-1—Using the circuit of Fig. 7-3 with $R_1 = R_2 = R_3 = 5K$, select values for C_1, C_2, and C_3 to make the oscillator frequency about 1 kHz.

SOLUTION—

$$C = \frac{1}{18\,Rf}$$

$$= \frac{1}{18 \times (5 \times 10^3) \times 10^3} = .011\ \mu F$$

In the final circuit, you could use 0.01-μF capacitors and a variable resistor for R_2 to adjust the frequency to 1 kHz.

TWIN T OSCILLATOR

Another type of RC oscillator that works well at audio frequencies is the *twin T* oscillator shown in Fig. 7-4. As in the phase-shift oscillator, this circuit uses an amplifier stage (Q2), and a phase-shifting network from the output of the amplifier stage back to the input. The phase-shifting network here consists of a high-pass T network in parallel with a low-pass T network, hence the name *twin T*. At a critical

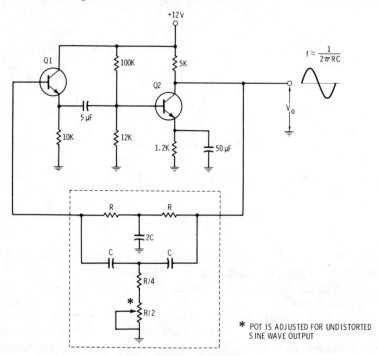

Fig. 7-4. Twin T oscillator.

frequency, where the output from the low-pass circuit begins to drop and the output from the high-pass circuit begins to rise, there exists a notch. At this notch frequency, the phase shift between the input to the filter (the collector of Q2) and the output from the filter (input to Q1) is 180°, and oscillation can occur. The pot in the center leg of the high-pass circuit is adjusted to make the circuit oscillate.

Transistor Q1 is used as an emitter follower to present a high impedance to the filter without losing signal available to the input of Q2.

An approximate frequency of oscillation can be found by the formula

$$f = \frac{1}{2\pi RC} \quad \text{(Eq. 7-1)}$$

where the R and C values are those labeled in the T networks. Note that the capacitor in the center leg of the low-pass T is twice the value of the other capacitors. Values of R should not exceed about 50K.

An excellent sine-wave oscillator can be made using an op amp as the amplifier, as shown in Fig. 7-5. Equation 7-1 also gives the approximate frequency of oscillation of the circuit of Fig. 7-5. Of course, if you want to be able to change the oscillator frequency, you can switch in different values of R and C.

EXAMPLE 7-2—Using the circuit of Fig. 7-5 with R = 50K, select values for the capacitors to make the circuit oscillate at a frequency of 1 kHz.

SOLUTION—Using equation 7-1,

$$C = \frac{1}{2\pi Rf} = \frac{1}{6.28 \times (5 \times 10^4) \times 10^3} = .00318 \ \mu F$$

In the final circuit you could use a .003-μF capacitor for each of the two values of C and twice that value for the 2C capacitor. Then

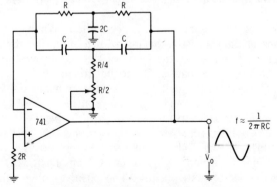

Fig. 7-5. Twin T oscillator using an op amp.

adjust the pot (25K in this case) to make the circuit oscillate. The resistor in series with the pot is not critical; a 10K resistor will give an adequate range of adjustment.

SQUARE-WAVE GENERATOR

Another waveform that is useful in testing audio amplifiers, and indispensible in digital work, is the square wave. Fig. 7-6 shows how a square-wave generator is constructed.

Recall that common emitter amplifiers have a 180° phase shift between the input and output. In Fig. 7-6 the output of Q1 is 180° out of phase with the input and is fed to the input of Q2. Since Q2 introduces another phase shift of 180°, the output of Q2 is *in phase* with the input of Q1. If the output of Q2 is connected to the input of Q1, oscillation can occur. However, this circuit does not produce sine waves. Instead, biasing resistors R_1 and R_2 are made small enough to cause the transistors to go into saturation, producing a square-wave output.

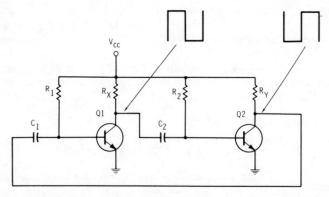

Fig. 7-6. Two cross-coupled amplifier stages form a square wave generator.

Here is how the square wave is generated. When power is first applied, one of the transistors begins to turn on harder than the other due to slight differences in the betas of the two transistors. Suppose Q1 turns on first. This causes Q2 to be driven off by the negative-going signal coupled through C_2. Q1 then remains on for awhile. Now as C_2 charges up through R_2, voltage at the base of Q2 becomes positive enough to begin to turn Q2 on. As Q2 begins to turn on, its collector voltage falls from $+V_{CC}$ toward ground (Q2 goes from cutoff to saturation). As the collector of Q2 gets less positive, a signal is coupled through C_1 which begins to turn Q1 off. Then as Q1 begins to turn off, its collector goes positive. The positive-going signal is coupled through C_2 to turn Q2 on harder. So the net effect is that Q2 is turned on hard, and Q1 becomes cut off. This is just the opposite of what happened when the power was first applied.

But remember, Q1 is cut off only by the charge on C_1. As C_1 discharges, the base of Q1 gets less and less negative until Q1 again begins to turn on. As Q1 turns on, a negative-going voltage is coupled through C_2 and again begins to turn off Q2. In other words, *each of the transistors is alternately cut off by the charge on the coupling capacitors.* Since C_1 charges through R_1, and C_2 charges through R_2, the time that each transistor is cut off is determined by the time constants R_1C_1 and R_2C_2.

The output of the circuit is taken at either collector and is a square wave with a peak-to-peak amplitude equal to V_{CC}. The frequency of the square wave can be approximately determined by the formula

$$f = \frac{1}{1.4\,RC}$$

where,
$R = R_1 = R_2$,
$C = C_1 = C_2$.

The square-wave generator, also called an *astable multivibrator,* is usually drawn as shown in Fig. 7-7.

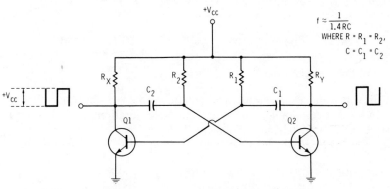

Fig. 7-7. An astable multivibrator (see also Fig. 7-6).

Almost any type of transistor can be used to build the square-wave generator. Collector resistors R_x and R_y are not critical; almost any value from a few hundred ohms to about 10K or 20K ohms can be used. For higher frequencies, you should use lower values of collector resistors, say 2K or less.

Resistors R_1 and R_2 must not be larger than beta times the collector resistor. For example, if you use a 2K collector resistor and the beta of the transistor is about 50, the values of R_1 and R_2 should not exceed

$50 \times 2K = 100K$. Smaller values of R can be used with no ill effects, but not so small as to cause excessive base current to flow.

Summarizing a design procedure for the oscillator of Figs. 7-6 and 7-7:

1. Choose a power supply voltage equal to the desired peak-to-peak output voltage (say 3 to 20 Vdc).
2. Choose $R_x = R_y = 500$ ohms to 20K (smaller R's for higher frequencies).
3. Choose $R_1 = R_2 < \beta R_x$.
4. Calculate the capacitor values, using equation 7-2; that is,

$$C = \frac{1}{1.4 \, Rf}$$

EXAMPLE 7-3—Design a square-wave generator with a 12-volt peak-to-peak output voltage and a frequency of 1 kHz. Assume the β's of the two transistors are both 30.

SOLUTION—
1. $V_{cc} \cong 12$ Vdc.
2. Let $R_x = R_y = 2K$.
3. Choose $R_1 = R_2 < \beta R_x < 30 \times 2K = 60K$. (Choose the closest standard value, say 56K.)
4. Solving for C,

$$C = \frac{1}{1.4 \, R_1 f} = \frac{1}{1.4 \times (5.6 \times 10^4) \times 1 \times 10^3} = 0.013 \, \mu F$$

The final circuit is shown in Fig. 7-8.

If you want a nonsymmetrical square wave, you can use different values for the coupling capacitors. The time duration that Q1 is held at cutoff is approximately equal to $0.7 R_2 C_2$. For example, using C_2

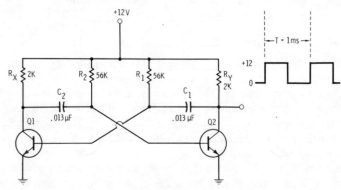

Fig. 7-8. A 1-kHz square wave generator.

equal to twice the value of C_1, Q2 will be cut off twice as long as Q1; thus the waveform at the collector of Q2 will be positive twice as long as it is zero. Fig. 7-9 shows the circuit of Fig. 7-8 revised to give a nonsymmetrical output. Of course, the waveform could also be made nonsymmetrical by using different values for R_1 and R_2, instead of different capacitor values. The important feature in adjusting the *duty cycle,* or percentage of on/off time of the signal, is that the duty cycle will be greater than 50% whenever $R_2C_2 > R_1C_1$, and less than 50% whenever $R_2C_2 < R_1C_1$.

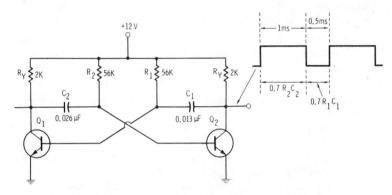

Fig. 7-9. Astable multivibrator with nonsymmetrical output.

One last modification can be made to allow you to vary the square wave frequency and duty cycle over a small range. If, instead of returning the upper ends of R_1 and R_2 to the V_{CC} supply, they are connected to a slider on a pot, the time that each transistor is held in cutoff can be varied by simply varying the pot (see Fig. 7-10).

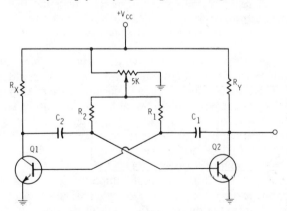

Fig. 7-10. Square wave generator with variable frequency and duty cycle.

135

Quiz

1. Using the circuit of Fig. 7-3 with $C_1 = C_2 = C_3 = 0.01$ μF, select values for R_1, R_2, and R_3 to set the oscillator frequency at about 555 Hz.
2. Suppose, in the twin T circuit of Fig. 7-5, $C = 0.01$ μF. Calculate the value of R which will give a frequency of about 1600 Hz.
3. Design a square-wave generator with a 9-V p-p output voltage and a frequency of about 715 Hz. Let $R_x = R_y = 10K$, and assume both transistors have a β of 50.
4. If you want to design the square-wave generator of problem 3 for a 10% duty cycle, what changes would be needed?
5. Estimate the dc current required for the square-wave generator of problem 3.

CONVERTING SINE WAVES TO SQUARE WAVES

A very simple way to produce a square wave is to saturate a high-gain amplifier. If a high-gain amplifier is driven with too large a sine-wave signal, the output will be driven from one power supply voltage to the other, clipping off the tops and bottoms of the sine wave; the output will be essentially a square wave. Fig. 7-11 shows an op amp

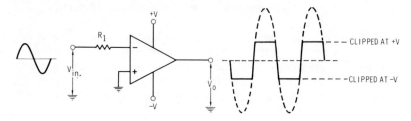

Fig. 7-11. A sine wave to square wave converter.

with the input driven by a sine wave of fairly large amplitude, say at least a volt or so. Notice that no feedback resistor is used from output to input. The amplifier gain is then equal to the open-loop gain of the device, which is extremely high. So, although the output tends to produce a sine wave of extremely large amplitude, most of the wave is clipped off, leaving essentially a square wave. Resistor R_1 is used to prevent excessive loading on the driving stage and also to prevent damage to the input of the converter.

If it is desirable to limit the output amplitude of the square wave to a lower voltage (say ±5 volts), two zener diodes can be connected back-to-back from the output to the input of the op amp, as shown in Fig. 7-12. (Zener diodes are discussed in Chapter 9.) A lower peak-

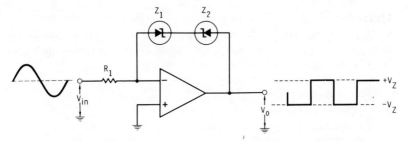

Fig. 7-12. Limiting the square wave amplitude with zener diodes.

to-peak voltage can also be obtained by reducing ±V of the op amp power supply to the desired level, if this is more convenient. However, zener diodes provide a more stable output.

CONVERTING SQUARE WAVES TO TRIANGULAR WAVES

If a square wave is fed to the input of an integrator circuit, the output from the integrator will be a triangular wave. As explained previously in Chapter 5, the output voltage of an integrator is given by

$$V_o = \frac{1}{R_1 C_1} \int_{t_o}^{t} V_{in} dt$$

If a constant amplitude voltage is fed to the input, the output will rise linearly.

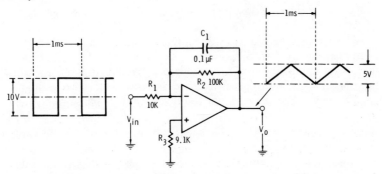

Fig. 7-13. A square wave to triangle wave converter.

In the circuit of Fig. 7-13, the input of the integrator is fed with a 1-kHz square wave. Notice that when the input is at a negative peak, say −5 volts, the output rises linearly. Then when the input switches to a positive peak, say +5 volts, the output falls linearly, thus developing a triangular wave. The amplitude of the triangular wave can be adjusted by varying either R_1 or C_1, but of course the maximum amplitude is limited by the op amp power supply.

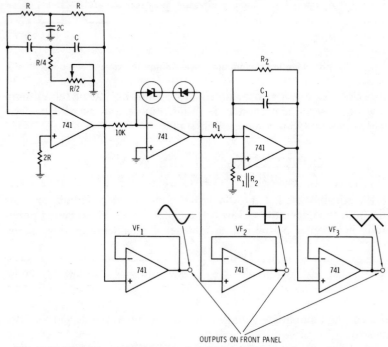

Fig. 7-14. Waveform generator produces sine, square, and triangle waves.

As mentioned before, resistor R_2 is needed to prevent the capacitor from charging up due to bias currents. The value of R_2 will determine the lowest frequency that the circuit can handle. If R_2 is too small, the output will simply be a distorted square wave rather than a triangular wave. The lowest input frequency should therefore be

$$f_{in} \geq \frac{1}{2\pi R_2 C_1} \qquad \text{(Eq. 7-3)}$$

EXAMPLE 7-14—What is the lowest frequency square wave that can be applied to the circuit of Fig. 7-13, to produce an undistorted triangular wave output?

SOLUTION—From equation 7-3,

$$f = \frac{0.159}{10^5 \times 10^{-7}} = 15.9 \text{ Hz}$$

WAVEFORM GENERATORS

In recent years, several manufacturers have been developing waveform generator which can produce sine waves, square waves, and triangular waves. With the circuits discussed in this chapter, you can design your own waveform generator to produce all three types of waves.

The circuit of Fig. 7-14 shows a complete waveform generator for a single frequency. The sine-wave generator is a twin T oscillator, as in Fig. 7-5. The output of the sine-wave generator is fed to a voltage follower (VF_1) and then connected to terminal posts on the front panel of the instrument. The voltage follower allows you to drive other circuits without loading the twin T excessively.

The output of the sine-wave generator is also fed to a square-wave converter, like the circuit of Fig. 7-12. The square-wave output is connected to the front panel through another voltage follower (VF_2) and also to the input of an integrator (see Fig. 7-13) to produce triangular waves. The triangular wave is also fed to the front panel through a third voltage follower, VF_3.

To make the waveform generator produce various frequencies, you will have to switch in different values of R and C for the twin-T circuit, and possibly different values of C for the integrator. Some experimentation with these values might be necessary to get the exact frequencies and amplitudes desired. Of course, the frequency of the square wave and triangular wave will always be the same as the sine-wave frequency.

The overall generator can be relatively small and inexpensive, since you can purchase dual 741 op amps, such as the type 747 or type 558, on a single chip at low cost. The entire circuit, then, can be built with just three chips. The overall size would be determined by how many different frequencies you want to generate, since a range of values is required for R and C.

SUMMARY

In this chapter we discussed some types of RC oscillators that are very useful at audio (and slightly higher) frequencies. Two sine-wave oscillators using discrete components were shown: the phase-shift oscillator and the twin T oscillator. The astable multivibrator, or square-wave generator, was also discussed.

We saw some applications for the op amp as a twin T sine-wave oscillator, a sine-to-square wave converter, and finally a square-to-triangular wave converter. These circuits are shown in Fig. 7-15.

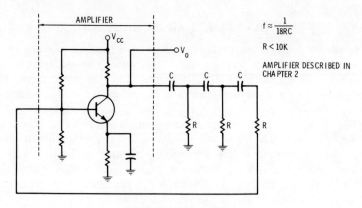

(A) Phase shift oscillator.

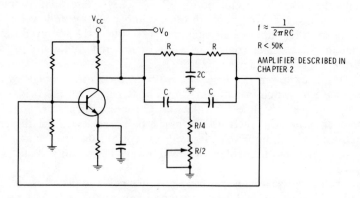

(B) Twin T oscillator.

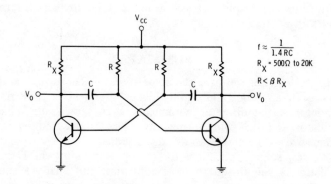

(C) Square wave generator.

Fig. 7-15. Summary

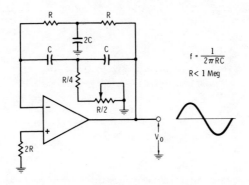

(D) Op amp twin T oscillator.

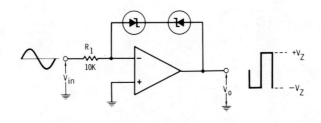

(E) Sine to square wave converter.

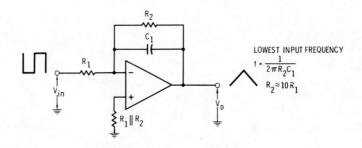

(F) Square to triangle wave converter.

of waveform-generating circuits.

8

Audio Power Amplifiers

Most audio power amplifiers in the past were designed as class-B or class-AB push-pull circuits, using an output transformer to couple the output stage to the speaker. In recent years, due to the increased availability of economical, complementary power transistors, the output transformer has been virtually eliminated. (Complementary transistors are pairs of npn and pnp transistors with similar characteristics.) In this chapter, we will discuss the design of a reliable form of complementary-symmetry amplifier.

COMPLEMENTARY-SYMMETRY AMPLIFIERS

Fig. 8-1 shows a simplified output stage using complementary symmetry. As shown, resistors R_1 and R_2 form a voltage divider across the power supply, giving a voltage at point B of 10 Vdc. When the power is first applied, output capacitor C_1 charges through Q1 to about 10 volts. When the voltage at point E reaches 10 Vdc, Q1 cuts off since its base is held at 10 volts. (The base of an npn transistor must be positive with respect to the emitter in order to bias the transistor on.) This is the quiescent state of the amplifier; both Q1 and Q2 are held at cutoff. This is called class-B bias.

Now what happens when an input signal is applied? As shown in Fig. 8-2A, when a positive-going signal is applied to the input, Q1 becomes forward-biased and turns on. Current flows through Q1, charging capacitor C_1 to a higher potential. Notice that Q1 acts like an emitter follower, feeding the speaker through C_1; so actually the ac signal appearing across the speaker is approximately the same as the input signal. Notice also that, on the positive half-cycle, Q2 is still cut off. However, on the negative half-cycle, the conditions are reversed: Q2 turns on as an emitter follower, and Q1 is cut off. When Q2 turns on, C_1 discharges through the speaker as shown in Fig. 8-2B.

The net result is that Q1 and Q2 conduct on alternate half-cycles, causing an ac current to flow through the speaker. Thus we have push-pull operation without using a transformer.

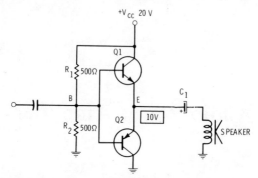

Fig. 8-1. Output stage using complementary symmetry.

The circuit of Fig. 8-1 works, after a fashion, but the output signal is rather distorted. The reason for this can be seen in Fig. 8-3. Assuming that we are using silicon transistors, neither transistor will conduct until its base is forward-biased by about 0.7 V. This means that Q1 will not actually conduct until the input voltage has gone positive by about 0.7 V. Similarly, Q2 will not conduct until the input voltage has gone negative by about 0.7 V. As a result, for a sinusoidal input like that of Fig. 8-3A, the output voltage across the speaker looks like Fig. 8-3B. The wrinkle in the waveform as it crosses the zero axis is called *crossover distortion,* and this crossover distortion introduces odd harmonics into the output signal.

In order to reduce crossover distortion, both transistors must be slightly forward-biased. For good efficiency we do not want to turn the transistors on very hard; so we will forward-bias each of them almost to conduction. Then even a slight positive or negative signal will cause conduction.

One simple way of forward-biasing both transistors is shown in Fig. 8-4. Resistor R_3 is placed between the two bases and adjusted so that there is about 1 volt across it. This way, both transistors are turned on slightly.

Although this method is sometimes used, it has the disadvantage of adding resistance in series with the input to Q2, resulting in slightly less signal applied to Q2 than to Q1.

Another way to remedy the situation is to use a pair of diodes between the two bases, as shown in Fig. 8-5. This has the advantage that negligible ac resistance appears in series with the input to the base of Q2. However, this circuit also has a major disadvantage: it is possible that the voltage drop across the two diodes in series might be so large that both transistors would be turned on hard. Not only is this condition inefficient, but the transistors might be turned on enough to be damaged.

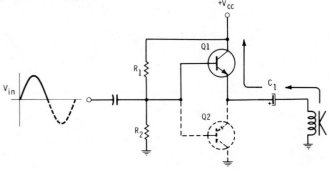

(A) C_1 charges through Q1 and speaker on positive half-cycle.

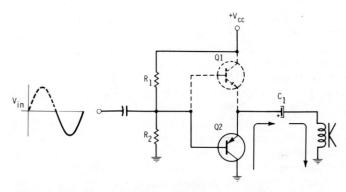

(B) C_1 discharges through Q2 and speaker on negative half-cycle.

Fig. 8-2. Complementary transistors function as emitter followers.

One of the best ways to forward-bias power transistors slightly is to use another transistor as shown in Fig. 8-6. In this circuit, resistors R_3 and R_4 turn Q3 on just enough so that the voltage dropped across it biases Q1 and Q2 the right amount. Transistor Q3 acts somewhat like an adjustable diode. The resistance values shown should work for most amplifiers. They can also be adjusted or "trimmed" during testing to minimize crossover.

Both power transistors should have about 5 to 10 mA of quiescent collector current flowing for efficient operation and negligible distortion.

Now consider the entire amplifier. Fig. 8-7 shows the complete diagram for a complementary-symmetry audio amplifier. The functions of Q3, Q4, and Q5 have already been discussed. Transistor Q2 is the driver transistor which feeds the ac signal to the power stage, and Q1 is the predriver, or voltage amplifier. The ac signal path is shown in heavy lines.

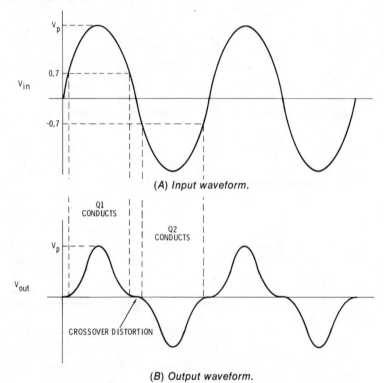

(A) Input waveform.

(B) Output waveform.

Fig. 8-3. Crossover disortion caused by transistor cutoff voltages.

Notice that for dc biasing, Q2 replaces resistor R1 of Fig. 8-6. The amount of current that must flow through Q2 is determined by the desired output power and load resistance. That is, Q2 must deliver the required base current for Q4 and Q5, which must be equal to I_{C4}/β_4. Likewise, the base current for Q2 comes from the collector of Q1. An example will illustrate how the various currents are established.

Suppose the power supply operates at 20 volts as indicated in Fig. 8-7. Resistors R_1 and R_2 form a voltage divider across the power supply so that the voltage at the base of Q1 is about 1.7 Vdc. This positive voltage at the base forward-biases Q1. As Q1 turns on, cur-

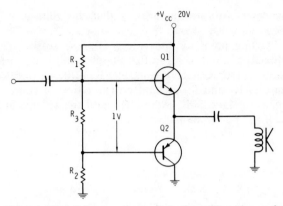

Fig. 8-4. Minimizing crossover distortion with a bias resistor.

rent flows into the base of Q2, activating it. As Q2 conducts more and more, the drop across it decreases, pulling the base of Q4 more positive. This causes Q4 to conduct more and more, pulling its emitter, and likewise point X, toward the base voltage.

Notice that resistors R_3 and R_4 form a 10-to-1 voltage divider from point X to ground. As the voltage at point X reaches about 10 volts (half the power supply), the voltage at the emitter of Q1 (point Y) reaches about 1 volt. This leaves 0.7 volts from base to emitter of Q1. If point X tries to go higher than 10 volts, the emitter of Q1 will be pulled more positive, thus reducing the forward bias on Q1 and tending to reduce its conduction. As a result, the voltage at point X never exceeds 10 volts ($\frac{1}{2}$ V_{CC}), which is just where we want it to remain.

A little thought will show that this circuit is also quite temperature-stable due to the negative feedback through R_3 and R_4. Regardless of the leakage currents, or values of beta of the various transistors, the

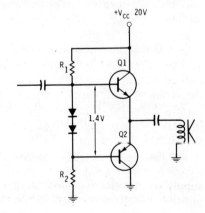

Fig. 8-5. Diode method of biasing.

operating currents will stabilize out so that the voltage at point X stays at about 10 volts.

A good starting point for the values of these resistors is to use 100 ohms for R_3 and adjust R_4 so that about 1 volt is dropped across R_3. In the above case, R_3 is 100 ohms, and R_4 is then about 900 ohms. Biasing resistors R_1 and R_2 are not critical either; a good value for R_2 is between 20K and 30K. Adjust R_1 until the voltage at point X measures half the power supply voltage.

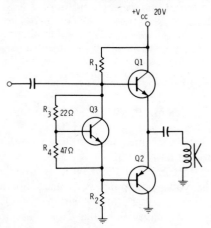

Fig. 8-6. Q3 acts as a variable biasing diode.

The two 0.47-ohm resistors in the emitters of the power transistors help reduce distortion; any value up to about 1 ohm is satisfactory here. Lastly, resistor R_7 must drop about half the power supply voltage with I_{C2} flowing through it.

In beginning the actual design, you must first decide how much output power you want and what speaker impedance you will use. Then you can proceed to calculate the required power supply voltage and circuit components. An example will help in explaining the actual design.

EXAMPLE 8-1—Design an audio power amplifier to work into a 16Ω speaker and having an output power of 5 W.

SOLUTION—First find the maximum current needed from each transistor. To develop 5 W in 16Ω, the required ac current is

$$I = \sqrt{\frac{P}{R_L}} = \sqrt{\frac{5}{16}} = \sqrt{.312} = 560 \text{ mA (rms)},$$

so the peak collector current is $I_{C\ max} = 1.41 \times I_{rms} = 1.41 \times 560$ mA $= 790$ mA.

Next, assuming that we can drive the power transistors almost to saturation, we can determine the required power supply voltage. To develop a peak current of 790 mA through 16Ω, the peak voltage across it must be

$$V_{max} = I_{c\ max} \times R_L = .790 \times 16 = 12.6V$$

The dc supply voltage must be twice this value, or 25.2V. The supply voltage must actually be a couple of volts or more higher than this to allow for some drop across each transistor; let the final working value of V_{cc} be about 30 Vdc.

The collector current of Q2 must be at least equal to the maximum required base drive current of the power transistors. (Only one power transistor will be turned on at a time.) The base drive for each power transistor will be equal to $I_{c\ max}/\beta_4$. Assuming a minimum β of 40, the maximum base drive current should be $I_{c2} = 790$ mA/$40 = 19.8 \cong 20$ mA. We should increase this value somewhat to prevent clipping, so let's use 1.5 times this value, or *30 mA*.

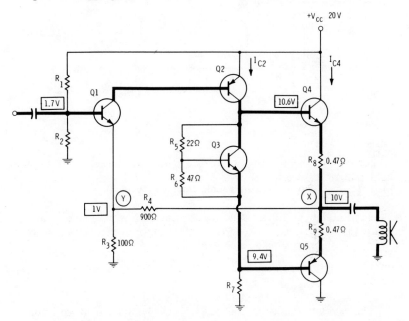

Fig. 8-7. Complete amplifier using complementary symmetry.

The current I_{c1} through Q1 must be equal to I_{c2}/β_2, which will probably be less than 1 mA. Actually, the value of I_{c1} will automatically adjust to the correct value because of the dc feedback caused by resistors R_3 and R_4, as explained previously. Using $R_3 = 100Ω$ with 1 volt across it as suggested, the value of R_4 should be large enough to drop the remaining 14 volts from point X, or

$$R_4 = 100\left(\frac{V_{cc}}{2} - 1\right) = 100\left(\frac{30}{2} - 1\right) = \mathit{1400Ω}$$

Now calculate the value of R_7. We must drop approximately $\frac{1}{2}V_{cc}$ across it with I_{C2} flowing through it. Since $I_{C2} = 30$ mA, we calculate the value of R_7 to be

$$R_7 = \frac{\frac{1}{2}V_{cc}}{I_{C2}} = \frac{15V}{30\text{ mA}} = 500 \text{ }\Omega$$

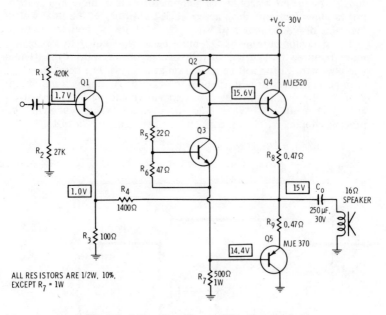

Fig. 8-8. A 5-watt audio amplifier (see text).

We can use about 20K to 30K for R_2; use 27K as a common standard value. Next, adjust R_1 to make the voltage at point X equal to 15 volts. An approximate value for R_1 is

$$R_1 \cong \frac{R_2 (V_{cc} - 1.7)}{1.7} \cong \frac{27K (30 - 1.7)}{1.7} \cong 450K$$

Use an approximate standard value such as *420K*.

The values of R_5, R_6, R_8, and R_9 given in Fig. 8-7 should work for almost any amplifier.

The complete amplifier is shown in Fig. 8-8. Notice the large value of C_o. The output capacitor must be large enough so that its reactance is about equal to the speaker impedance at the lowest frequency that the amplifier must handle. In this case, C_o has about 16 ohms of capacitive reactance at a frequency of 40 Hz.

According to class-B amplifier theory, the power dissipation of each power transistor should be about 0.25 times the output power, or about 1.25 watts for this 5-W amp. Actually, the amplifier is operating class AB, so the power dissipation will be slightly higher. Suitable power transistors must be capable of at least this much (1.25 W) power dissipation. In choosing transistors for Q4 and Q5, complementary pairs of transistors should be used, such as MJE520 and MJE370, or TIP29 and TIP30, etc.

The driver transistor, Q2, is operating class A, so its power dissipation can be calculated by the product of its dc collector voltage times its dc collector current. In this case, the power dissipation of Q2 is $P = 15 \times .03 = 0.45$ watt. Transistors Q1 and Q3 dissipate a negligible amount of power, and these two need not be power transistors.

After this somewhat lengthy discussion, it should prove helpful to summarize a design procedure.

Design Summary for the Amplifier of Fig. 8-7

Use the values shown for resistors R_5, R_6, R_8, and R_9. P = desired output power in watts (rms); R_L = speaker impedance in ohms.

1. Find $I_{C\ max}$ for power transistors.

$$I_{C\ max} = I_{L\ max} = 1.41 \times \sqrt{\frac{P}{R_L}}$$

2. Calculate power supply voltage.

$$V_{CC} = 2 \times I_{C\ max} \times R_L$$

3. Use $R_3 = 100$ ohms.

4. $R_4 = 100 \times \left(\frac{V_{CC}}{2} - 1\right)$.

5. Calculate driver transistor collector current.

$$I_{C2} \cong 1.5 \times \frac{I_{L\ max}}{\beta_4}$$

(β_4 is the β of Q4)

6. Next calculate the value for R_7 according to

$$R_7 = \frac{\frac{1}{2} V_{CC}}{I_{C2}}$$

7. $R_2 - 20K$ to $30K$.

8. The value of resistor R_1 is given by

$$R_1 = \frac{R_2(V_{CC} - 1.7)}{1.7}$$

9. The minimum value of coupling capacitor C_o is governed by the minimum frequency applied (f_{min}), or

$$C_o = \frac{1}{2\pi R_L f_{min}}$$

(for −3 dB at f_{min})

10. Calculate the power dissipated by Q4 and Q5.

$$P_{C5} = P_{C4} = 0.25 \times P_L$$

11. Calculate the power dissipated by Q2.

$$P_{C2} = \frac{V_{CC}}{2} \times I_{C2}$$

Quiz

1. What maximum rms power will the amplifier of Fig. 8-8 develop if the output is connected to an 8-ohm speaker? Assume other factors have been adjusted so that the rms current through each transistor remains constant at 560 mA maximum.
2. Suppose no "other factors" had been adjusted in problem 1. (A) How much rms power would then be developed with an 8-ohm speaker? (B) Does this exceed the design limits of the amplifier? (C) If so, what can be done if you want to operate the amplifier with 8-ohm speakers?

HEAT SINKS

When working with power semiconductors having a dissipation of more than about 1 watt, heat sinks should be used. Even if the device is not actually destroyed by overheating, the operating life will be shortened by thermal fatigue if it is required to operate at excessively high temperatures. We will now discuss some factors in selecting an appropriate heat sink.

The purpose of a heat sink is to transfer the heat generated by a device and dissipate it over a large area so that the air can carry it away. For large amounts of power where space is restricted, forced air or even liquid cooling may be necessary. Most of the time, however, ordinary convection is adequate.

Heat generated at the junction of a transistor must travel out to the case in order for it to be carried away. This heat flow is a rather slow process, generally, and the opposition to the outward flow of heat is called the *thermal resistance* of the device. The *thermal resistance from junction to case,* designated θ_{JC}, is specified in °C/watt and is given by the manufacturers of the power devices. A typical value for a silicon power transistor in a TO-3 case is 1.5°C/W. Values may actually range from 1 to 50°C/W.

When mounted on a heat sink, a transistor is sometimes insulated from the metal sink by a washer and some silicone grease. The washer also has a thermal resistance, θ_{CS}, which specifies the thermal resistance from case to sink. If unknown, this value can be approximated at about 0.5°C/W.

Finally, there is the thermal resistance from sink to air, θ_{SA}, which is determined by factors such as the size and shape of the heat sink. Manufacturers of heat sinks list values of thermal resistance for the heat sinks they make. The required value should be calculated, or at least approximated, before you choose an appropriate heat sink. Generally, a larger surface area for a heat sink implies a smaller thermal resistance and, consequently, less temperature rise for a given power dissipation.

The following equation relates power dissipation to junction temperature, ambient temperature, and thermal resistance:

$$P = \frac{T_J - T_A}{\theta_{JC} + \theta_{CS} + \theta_{SA}} \qquad \text{(Eq. 8-1)}$$

where,

P is the power dissipation in watts,
T_J is the maximum allowable junction temperature in °C,
T_A is the maximum ambient temperature in °C,
θ_{JC} is the thermal resistance from junction to case,
θ_{CS} is the thermal resistance from case to sink,
θ_{SA} is the thermal resistance from sink to air.

EXAMPLE 8-2—You are using a transistor with a thermal resistance θ_{JC} of 1.5°C/W, and the junction temperature is not to exceed 125°C. Assume 0.5°C/W for θ_{CS}. The maximum ambient temperature is 50°C, and the transistor is to dissipate 15 watts. Calculate a suitable value for the heat-sink thermal resistance, θ_{SA}.

SOLUTION—From equation 8-1,

$$\theta_{SA} = \frac{T_J - T_A}{P} - \theta_{JC} - \theta_{CS}$$
$$= \frac{125 - 50}{15} - 1.5 - 0.5$$
$$= 5 - 1.5 - 0.5$$
$$= 3°C/W$$

Any heat sink with a thermal resistance of 3°C/W *or less* could be used. The final choice would be one of cost and space.

Manufacturers often specify the maximum *case* temperature of transistors, rather than the maximum junction temperature. In such instances, the thermal resistance from junction to case need not be specified. The equation for calculating the appropriate heat sink thermal resistance then becomes

$$\theta_{SA} = \frac{T_C - T_A}{P} - \theta_{CS} \qquad \text{(Eq. 8-2)}$$

where T_C is the maximum allowable case temperature, and all other quantities are as before.

Of course if the transistor is mounted directly to the metal sink without any insulation, θ_{CS} can be omitted. Be sure sufficient ventilation of the heat sink is provided; usually this can be ensured simply by mounting the heat sink on the outside of the case, in a position where it will always be exposed to ambient air.

Quiz

3. The junction temperature of a transistor is not to exceed 150°C and θ_{JC} is specified to be 2.0°C/W. If the maximum ambient temperature is 50°C, calculate the maximum θ_{SA} required for a power dissipation of 10 watts. Assume that $\theta_{CS} = 0.5°C/W$.
4. If the transistor in problem 3 need not be electrically insulated from the heat sink, what will the new value of θ_{SA} be?
5. A temperature of 50°C corresponds to about 123°F. Is this a safe maximum temperature to use for all applications? If not, think of instances where a higher value might be needed.

POINTERS ON CONSTRUCTING POWER AMPLIFIERS

Power amplifiers sometimes have a tendency to oscillate or amplify extraneous signals. You can minimize this tendency as well as possible damage to the unit by following these construction hints.

1. If at all possible, use one common ground point. Run the common leads from each stage to one common point at the power supply. If this is not feasible, use a heavy ground bus and connect all ground leads to it.
2. It may be necessary to decouple the predriver stage. This is accomplished by splitting the biasing resistor and connecting a capacitor from the junction to ground (see Fig. 8-9). Decoupling prevents variations in power supply voltage from having as much effect on the input.
3. Also to prevent oscillation, it may be necessary to connect a small capacitor (say 0.02 μF) from the base of Q5 to ground. Another possible remedy is to connect a 0.05-μF capacitor in

series with a small resistor (about 27 ohms) across the speaker. Some experimenting in this regard may be necessary.
4. Use a well-filtered power supply.
5. Since the input impedance is high, it would be advisable to use shielded leads for the input.
6. Keep low-level leads short.
7. Keep the low-level stages physically away from the output stages and away from the power supply transformer.
8. Use overload protection in the power supply (see Chapter 9).
9. Transistors Q2, Q4, and Q5 may need heat sinks.
10. If it is necessary to insulate a transistor electrically from a heat sink, use an insulating wafer or washer, and coat both sides liberally with silicone grease for better heat conduction.

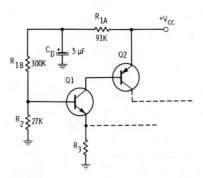

Fig. 8-9. Decoupling the bias circuit of Fig. 8-8.

SUMMARY

In this chapter we studied a complementary-symmetry audio power amplifier. Although there are various other designs of complementary-symmetry amplifiers, the one shown is relatively easy to design without using critical components and gives good results. The only components that require care in selection are the power transistors. You should use complementary pairs, such as MJE520 and MJE370, TIP-29 and TIP-30, etc. The complete power amplifier is shown in Fig. 8-7, and a condensed design summary is outlined in the text.

Whenever the power dissipation of a device exceeds about 1 watt, you should use a heat sink. If the maximum junction temperature of a semiconductor is given, use equation 8-1 to determine the heat sink thermal resistance. If the maximum case temperature of the device is given, use equation 8-2. Omit θ_{CS} from either equation if no insulating material is used between the device and the heat sink.

9

Regulated Power Supplies

From the previous discussion of power supplies in Chapter 1, you know how to design various types of supplies to give you a specified value of output voltage and current. However, will the output be constant even if the load current changes, or if the input line voltage changes? The answer is no. Whenever the load current or input voltage for a supply changes, the output voltage will vary somewhat. The more it varies, the more trouble it will cause. This is especially true in equipment such as test instruments or computer circuits; a change in power supply voltage can cause erroneous readings. In this chapter, we will look at some ways to build regulated power supplies—that is, power supplies with an output voltage that remains constant, even if the load current or input voltage varies.

For convenience in assigning semiconductor voltage drops, the devices referred to in this chapter are assumed to be silicon (0.7-V drop). Remember to modify this value when dealing with other types of semiconductors.

ZENER DIODE REGULATOR

One of the simplest types of voltage regulators is the *zener diode*. A zener diode is simply a silicon diode built to operate in reverse breakdown. Fig. 9-1 shows the circuit setup for determining the zener breakdown voltage. Starting with the power supply at zero volts, the voltage is gradually increased. If the zener diode is good, practically no current will flow in the reverse direction, as long as the voltage is not high enough to cause breakdown. However, as you learned in Chapter 1, if a high enough value of reverse voltage is applied across a diode, the diode will suddenly begin to conduct. This condition is called *breakdown*.

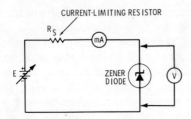

Fig. 9-1. Circuit for measuring zener diode breakdown voltage.

Breakdown does not mean that the diode is damaged. The diode will not be damaged as long as the power dissipation of the diode is held within the limits specified by the manufacturer. Here, breakdown refers instead to the minimum reverse voltage that will cause the zener to start conducting.

The voltage at which the diode goes into breakdown is called the zener voltage, V_z. Zener diodes come in a wide range of voltages, from a few volts to a few hundred volts. They are available in power ratings of 250 mW, 500 mW, 1 W, 10 W, 50 W, and a few others as well.

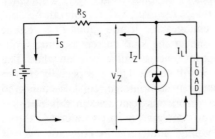

Fig. 9-2. Simple zener diode voltage regulator.

Here is how the zener regulator works (refer to Fig. 9-2). Resistor R_s is chosen so that it drops the difference voltage between E and V_z. The voltage across the load is, of course, equal to V_z. Notice that in Fig. 9-2, the two currents I_z and I_L join up to form I_s. If for any reason I_L decreases, the voltage across the load tends to increase. But as soon as the voltage across the zener tries to increase, even slightly, the current through the zener increases rapidly, holding the voltage across it essentially constant. In other words, as the load current decreases, the zener current increases and vice-versa. The load voltage will remain essentially equal to V_z even when the load current changes, as long as the zener remains in the breakdown region. If the input voltage changes, the load voltage will again tend to change. With a zener diode across the load, the load voltage will still remain essentially constant.

There are two extreme conditions that must be satisfied if the regulator is to work properly. First of all, there must be some current flowing through the zener, at least a couple of milliamps, to keep it in breakdown. Second, an excessive current flowing through it will burn it out. Keep these two points in mind in our design procedure.

Before designing the regulator, we must know (usually by measurement) what the maximum load current will be and also what the maximum and minimum values of input voltage will be.

To satisfy the first extreme condition, we must choose a value for R_s to drop the difference in voltage between the minimum value at E and voltage V_z. The current through R_s at this time cannot be any lower than the maximum current drawn by the load plus the minimum current through the zener to maintain breakdown (usually 1-5 mA). Mathematically, we find the required value of R_s to be

$$R_s = \frac{E_{min} - V_z}{I_{L\ max} + I_{z\ min}} \qquad \text{(Eq. 9-1)}$$

where,

R_s is the series resistance in ohms (Fig. 9-2),
E_{min} is the minimum power supply voltage,
V_z is the zener breakdown voltage,
$I_{L\ max}$ is the maximum load current in amps.
$I_{z\ min}$ is the minimum zener current in amps.

Next, to avoid burning out the zener, we must choose one with a sufficient power rating. To find the power rating of the zener, we must know what maximum current will flow through it. The largest value of zener current will flow if the load current is zero (load disconnected) and, at the same time, the input voltage is at the maximum value.

With E at a maximum, the current through R_s will increase to

$$I_{s\ max} = \frac{E_{max} - V_z}{R_s} \qquad \text{(Eq. 9-2)}$$

where,

$I_{s\ max}$ is the maximum current through R_s,
E_{max} is the maximum power supply voltage.

So, the maximum power dissipated in R_s is

$$P_{R\ max} = I^2_{s\ max} \times R_s \qquad \text{(Eq. 9-3)}$$

If I_L happens to be zero at the same time, then

$$I_{z\ max} = I_{s\ max}$$

Finally, the power dissipated in the zener is

$$P_{z\ max} = V_z \times I_{z\ max} \qquad \text{(Eq. 9-4)}$$

EXAMPLE 9-1—In a circuit like that of Fig. 9-2, I_L varies from 0 to 50 mA, $V_z = 12$ volts, and E varies from 15V to 17V. What value should be used for R_s? What is the maximum power dissipated in R_s? What is the maximum power dissipated in the zener? Assume $I_{z\ min} = 5$ mA.

SOLUTION—From equation 9-1,

$$R_s = \frac{E_{min} - V_z}{I_{L\ max} + I_{z\ min}} = \frac{15V - 12V}{50\ mA + 5\ mA} = \frac{3V}{55\ mA} = 54.5\Omega$$

From equation 9-2, the maximum series current will be

$$I_{s\ max} = \frac{E_{max} - V_z}{R_s} = \frac{17 - 12}{54.5} = \frac{5}{54.5} = 92\ mA$$

So the maximum power dissipated in R_s will be

$$P_{R\ max} = I^2_{s\ max} \times R_s = (92\ mA)^2 \times 54.5\Omega = 0.457\ W$$

And from equation 9-4,

$$P_{z\ max} = V_z \times I_{z\ max} = 12V \times 92\ mA = 1.1\ W$$

Summary of Zener Regulator Design

Before beginning the design of a zener regulator, you must decide on a few things. First, you must know what load voltage and load current will be needed. Next, you must know what the input voltage will be. That is, since the input voltage varies, determine both the maximum and minimum voltage values. The minimum value should be at least a couple of volts higher than the zener voltage, so that some voltage can be dropped across R_s. Assuming all the above factors are known (at least approximately), we can establish a design procedure.

1. Choose $V_z = V_L$ (the desired load voltage).
2. Calculate R_s from equation 9-1:

$$R_s = \frac{E_{min} - V_z}{I_{L\ max} + I_{z\ min}}$$

   ($I_{z\ min}$ can usually be taken as 5 mA)
3. Calculate the maximum series current with equation 9-2:

$$I_{s\ max} = \frac{E_{max} - V_z}{R_s}$$

4. Calculate the maximum power that will be dissipated in R_s using equation 9-3:

$$P_{R\ max} = I_{s\ max}^2 \times R_s$$

5. Finally, calculate the maximum power that will be dissipated in the zener using equation 9-4:

$$P_{z\ max} = V_z \times I_{z\ max}$$

HIGHER CURRENT REGULATORS

The zener diode regulator is economical and easy to use, but it has limitations. For one thing, the output current from a zener-regulated supply is not very high. Remember that the zener diode must conduct whatever current the load does not conduct. That is, the current through series resistor R_s is constant, whether or not the load draws current. So when the load current drops to the minimum value, the zener must conduct the remainder of the current. This means that for high current power supplies, say on the order of several amperes, the zener diode would have to be able to conduct this amount. Even if such a zener is available, the circuit will not be very efficient in terms of power consumption.

We will now discuss a more efficient way of regulating the power supply voltage. In the circuit of Fig. 9-3, we see a transistor used as an emitter follower. The output voltage, taken at the emitter, is essentially the same as the voltage from base to ground; that is, if $V_B = $ 10 volts, the voltage at the emitter is also approximately 10 volts. More exactly, if we consider the 0.7-volt drop across the base-emitter junction, the voltage at the emitter is 9.3 volts.

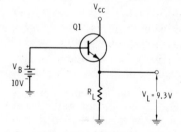

Fig. 9-3. Voltage across R_L is determined by emitter-base voltage of Q1.

Notice that we do not consider the value of R_L. This resistance makes no appreciable difference; the voltage at the emitter is still about 9.3 volts. The circuit of Fig. 9-3, then, provides a basis for an improved voltage regulator. Even though the load may draw different amounts of current, the voltage across it will be essentially equal to the voltage at the base of Q1.

Now we would like to dispense with the battery connected to the base. As shown in Fig. 9-4, we can replace this battery with a zener diode. Notice that the zener is held in conduction by current flowing through R_s as before. The base current for the transistor must also flow through R_s.

Here is the big advantage of this circuit over the simple zener regulator. When the load current is high, the high current flows through the transistor (usually a power transistor); but when the load current

161

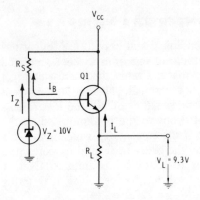

Fig. 9-4. Replacing the bias battery with a zener diode.

drops to a very low value, the zener does *not* have to carry and dissipate this large current. The current through the zener does increase, but only by an amount equal to the change in base current of the transistor. That is, the zener current increases by ΔI_L *divided by* the beta of the transistor, which is only a small fraction of the full load current, I_L.

The transistor used in this regulator is in series with the load, so the circuit is called a *series* regulator. The transistor is usually referred to as the *pass* element, since it passes current to the load.

Maximum power will be dissipated in the pass transistor when the input voltage is maximum and the load current is maximum. That is,

$$P_{C\ max} = (E_{max} - V_L) \times I_{L\ max} \qquad \text{(Eq. 9-5)}$$

where,
- $P_{C\ max}$ is the maximum power dissipated in watts,
- E_{max} is the maximum input (or power supply) voltage,
- V_L is the regulated load voltage,
- $I_{L\ max}$ is the maximum load current in amps.

The series regulator is usually drawn as shown in Fig. 9-5.

EXAMPLE 9-2—Design a series regulator like the one of Fig. 9-5, with an output voltage of about 5.5 Vdc and a maximum load-current capability of 500 mA. Assume that input voltage E varies from 9 to

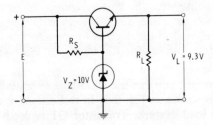

Fig. 9-5. Series regulator.

12 Vdc. Also assume that the transistor has a beta of 50, and let $I_{z\ min} = 5$ mA.

SOLUTION—First of all, since the load voltage is to be 5.5 volts, the voltage at the base of the transistor (hence the zener voltage) should be 0.7 volts higher, or 6.2 volts. Next, to calculate the value of R_s we must use the minimum input voltage of 9 volts. At this time, the current through R_s will be

$$I_{z\ min} + \frac{I_{L\ max}}{\beta} = 5\ mA + \frac{500\ mA}{50} = 15\ mA$$

So

$$R_s = \frac{E_{min} - V_z}{15\ mA} = \frac{9 - 6.2}{15\ mA} = 187\Omega \simeq 180\Omega$$

As before, the maximum zener current will be equal to the maximum current through R_s, or

$$I_{z\ max} = I_{s\ max} = \frac{E_{max} - V_z}{R_s} = \frac{12 - 6.2}{180} = 32\ mA$$

So the maximum power dissipated in the zener will be

$$P_z = V_z \times I_{z\ max} = 6.2 \times 32\ mA = .196\ W$$

A 250-mW (¼ W) zener will be adequate. The power dissipated in R_s will be

$$P_R = I_{s\ max}^2 \times R_s = (32\ mA)^2 \times 180\Omega = .185\ W$$

Use either a ¼-W or ½-W resistor.

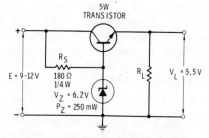

Fig. 9-6. Circuit for Example 9-2.

Finally, the maximum power dissipated in the transistor will be (equation 9-5)

$$P_c = (12 - 5.5) \times .5 = 3.25\ W$$

We should use at least a 5-W transistor. The final circuit is shown in Fig. 9-6.

For even greater output currents, you can use a Darlington arrangement for the pass transistor, as shown in Fig. 9-7. In this circuit, only transistor Q2 carries the full load current. Transistor Q1 is used to

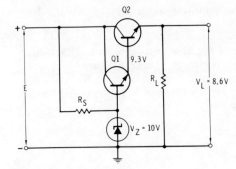

Fig. 9-7. Darlington arrangement provides high current capability.

feed the base of Q2, so the current through R_s will only be I_z plus the base current of Q1, which is very small compared to the load current.

We can now establish a design procedure for a higher-current power supply regulator like the one of Fig. 9-5. As in the simple zener-regulated supply, we must first know the desired load voltage, load current, and the maximum and minimum values of dc input voltage. If you are starting from scratch (that is, if you are going to build a rectifier power supply for the input dc voltage), you can use the procedures discussed in Chapter 1. A good starting point would be to choose a transformer with a *secondary rms* voltage equal to the dc voltage that you want out of the regulator. For example, if you need a 10-volt regulated supply, you can choose a transformer with a 10-volt *rms* secondary. This will give you about 14 volts dc across the filter capacitor, which will provide about 4 volts extra to be dropped across R_s. Also, if the line voltage varies, the dc voltage across the filter capacitor will vary. The line voltage will normally not vary more than about ±10%; so the dc voltage across the filter will also vary about ±10%. You can use this ±10% value to find E_{max} and E_{min}. For example, if you use the 10-volt rms transformer, the rectified voltage across the filter capacitor will be about 14 volts, ±1.4 volts, or $E_{min} = 12.6$ Vdc, and $E_{max} = 15.4$ Vdc.

Summary of Design Procedure for Higher-Current Regulated Supply

The following notes apply to the design of a voltage regulator like the one shown in Fig. 9-5.

1. Choose $V_z = V_L + 0.7$ V. (V_z must be slightly higher than V_L due to the transistor base-emitter voltage drop.)
2. Calculate R_s by

$$R_s = \frac{E_{min} - V_z}{I_{z\,min} + I_B/\beta}$$

3. Calculate the maximum series current through R_s by
$$I_{s\ max} = \frac{E_{max} - V_z}{R_s}$$
4. Calculate the maximum power that will be dissipated in R_s,
$$P_{R\ max} = I_{s\ max}^2 \times R_s$$
5. Calculate the maximum power that will be dissipated in the zener,
$$P_{z\ max} = V_z \times I_{z\ max}$$
6. Finally, calculate the maximum power that will be dissipated in the transistor.
$$P_C = (E_{max} - V_L) \times I_{L\ max}$$

Quiz

Choose the correct word or words in each of the following questions about basic regulators.

1. A zener diode regulator is used to maintain a constant load (voltage, current).
2. Simple zener diode regulators are best used for (high-, low-) current applications.
3. If resistor R_s in Fig. 9-2 is made too large, the zener diode will (burn up, come out of conduction) when E_{in} drops to the minimum value).
4. You should calculate the power rating of the zener diode when E is (max, min) and when I_L is (max, min).
5. You should calculate the resistance value of R_s when E is (max, min) and I_L is (max, min).
6. The pass transistor in Fig. 9-6 will dissipate maximum power when E is (max, min) and I_L is (max, min).
7. The output voltage of the regulator of Fig. 9-5 will always be (slightly higher, slightly lower) than V_z.

VARIABLE VOLTAGE REGULATORS

Thus far we have discussed only fixed-voltage regulators. Often, say for testing circuits, you will have need of a variable voltage regulator; that is, you will want to set the output voltage at some specified value and have it remain at that value during use. We will now modify the higher-current regulator to make it variable. Consider the circuit of Fig. 9-8. This circuit is essentially a noninverting amplifier. The output voltage, V_L, across load R_L is given approximately by

$$V_L = V_{ref} \times \frac{R_1 + R_F}{R_1}$$

Notice that the value of output voltage is essentially independent of R_L, as long as R_L is fairly large. Here we have the essence of a voltage regulator.

Fig. 9-9 is a slight modification of Fig. 9-8. Notice that resistors R_1 and R_F are replaced with potentiometer R_V. By varying the position of the pot slider, we can effectively change the ratio R_F to R_1, thus changing the gain of the circuit. Changing the circuit gain also changes output voltage V_L.

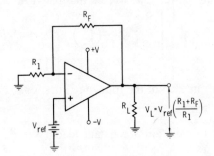

Fig. 9-8. Using an op amp to control load voltage V_L.

If the pot slider is at the extreme right position the circuit acts like a voltage follower; $V_L = V_{ref}$. Then as we move the slider more and more to the left, the output voltage gets higher and higher until eventually it hits the positive power supply voltage (saturation). *Thus, V_L is variable from V_{ref} to the $+V$ supply voltage.*

The circuit of Fig. 9-9 is essentially a variable voltage regulator. However, the op amp probably cannot deliver much output current, so we can now add another component to increase the output current capability.

Fig. 9-10A shows the same circuit as in Fig. 9-9, but drawn in a slightly different manner. Output voltage V_L can still be varied with

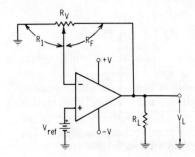

Fig. 9-9. V_L is variable by adjusting R_V.

R_V. Next, in the circuit of Fig. 9-10B, an emitter follower has been added after the op-amp output. You will remember that the output voltage of an emitter follower is essentially the same as the input voltage to the base. By putting R_V in the emitter lead, the op-amp circuit is still essentially the same. The only difference is that the load current through R_L comes from Q1 rather than from the op amp. The emitter follower, then, increases the current output capability of the regulator.

The emitter follower acts as the pass transistor and is usually drawn as shown in Fig. 9-11.

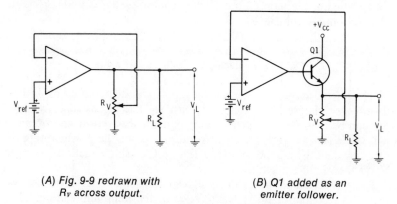

(A) Fig. 9-9 redrawn with R_V across output.

(B) Q1 added as an emitter follower.

Fig. 9-10. Increasing the op amp current capacity.

Next, we can eliminate the V_{ref} battery by replacing it with a zener diode and series resistor R_s, as shown in Fig. 9-12.

Finally, we need $+V$ and $-V$ connections for the op amp itself. The complete regulator is shown in Fig. 9-13. Resistors R_1 and R_2 are used to limit the output voltage variations. A good starting point would be to use a 1.2K resistor for both R_1 and R_2, and a 2.5K pot for R_V. These values can be changed during testing.

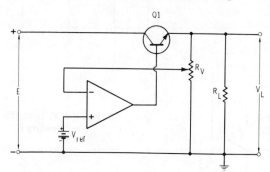

Fig. 9-11. Normal representation of Fig. 9-10B.

167

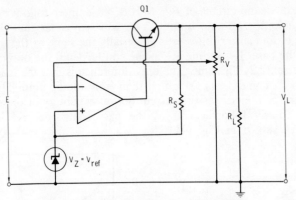

Fig. 9-12. Replacing V_{ref} with a zener diode.

The lowest output voltage for the supply will be about a volt higher than V_z; so choose V_z to suit your purpose. The highest that the output can be adjusted is approximately equal to E. Almost any op amp will work (a 741 gives excellent results). As for the power rating of Q1, calculate the maximum dissipation in Q1 using equation 9-5 with V_L at the minimum setting.

Resistor R_s should be small enough so that some zener current flows (at least a couple of mils) when V_L is at the minimum setting. This will give about a 1-volt drop across R_s. Usually, about 500Ω is a good starting value. Then be sure to determine the maximum current that will flow through R_s when V_L is set at the maximum value. That is,

$$I_{s\ max} = I_{z\ max} = \frac{V_{L\ max} - V_z}{R_s} \qquad \text{(Eq. 9-6)}$$

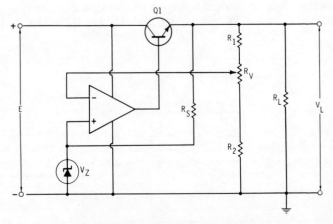

Fig. 9-13. Complete variable regulator.

Design Summary

The following procedures apply to the design of a voltage regulator like the one shown in Fig. 9-13.

1. Choose $V_z = V_{L\ min} - 1\ V$

2. Set $R_s = \dfrac{1\ V}{I_{z\ min}} \cong 500\Omega$

3. Choose $R_1 = R_2 = 1.2K$, $R_V = 2.5K$
4. Determine power dissipation of Q1 by

$$P_C = (E_{max} - V_{L\ min}) \times (I_{L\ max})$$

5. Determine $P_z = V_z \times I_{z\ max}$
6. Determine power rating of R_s by

$$P_R = I_{s\ max}^2 \times R_s$$

EXAMPLE 9-3—Design a regulator with an output adjustable from 4 V to 12 V. The maximum load current is to be 200 mA. Assume you already have an unregulated dc supply with an output that changes from 14 V to 16 V with line voltage variations.

SOLUTION—Using the circuit of Fig. 9-13, $V_z = 4 - 1 = 3$ V, let $R_s = 500\Omega$. The maximum power dissipation in the transistor will be

$$P_c = (16 - 4) \times 0.2 = 2.4\ W$$

Use a 5-W transistor. Then, from equation 9-6, we find the maximum zener current to be

$$I_{z\ max} = \dfrac{12 - 3}{500} = 18\ mA$$

So $P_z = 3\ V \times 18\ mA = 54\ mW$. Any 250-mW zener will work fine. Also $P_{Rs} = (18\ mA)^2 \times 500 = 0.162\ W$; a ¼-W to ½-W resistor will be more than adequate.

CURRENT LIMITERS

To protect power supplies from damage, you must limit the current drawn from them. Ordinary fuses cannot operate fast enough to keep the solid-state components from being ruined before the fuse opens. We will now look at a better way to prevent overload on a power supply.

Fig. 9-14 shows a simple current-limiter circuit. Resistor R_B is chosen such that the transistor is operating at saturation. Thus, the current through the load is determined by R_L, R_E, and supply voltage E. Resistor R_E is so small that the drop across it is normally less than 0.7V; hence, diodes X1 and X2 are not forward-biased.

If R_L decreases, the current through it increases. However, as R_L decreases, the drop across R_E increases due to the increased current. When the drop across R_E reaches 0.7 V, the total drop across the two diodes reaches 1.4 V, which causes them to become forward-biased. As the diodes begin to conduct they draw away some of the current through R_B, causing Q1 to conduct less and preventing the load current from increasing.

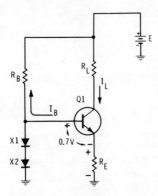

Fig. 9-14. Basic current limiter.

Putting it another way, the drop across both diodes in series cannot exceed 1.4 V, so the drop across R_E is prevented from exceeding 0.7 V. The load current through R_E is therefore limited to a value of $I_L = 0.7 \text{ V}/R_E$. By choosing the proper value of R_E, you can limit the load current to any value you choose.

EXAMPLE 9-4—In the circuit of Fig. 9-14, if $E = 14$ V and $\beta = 50$, choose values of R_B and R_E to limit the current to 200 mA.

SOLUTION—First of all, in order to saturate the transistor with up to 200 mA flowing through it, the base current through Q1 should be

$$I_B = \frac{I_c}{\beta} = \frac{200 \text{ mA}}{50} = 4 \text{ mA}$$

Resistor R_B will have a drop across it equal to supply voltage E minus the drop across the two diodes, or $14 - 1.4 = 12.6$ V.

$$R_B = \frac{12.6 \text{ V}}{4 \text{ mA}} = 3.15\text{K} \cong 3\text{K}$$

Then, to limit the current to 200 mA, the drop across R_E will reach a maximum value of 0.7 V, or

$$R_E = \frac{0.7 \text{ V}}{0.2 \text{ A}} = 3.5 \text{ }\Omega$$

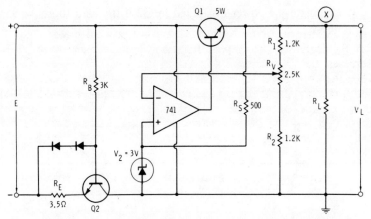

Fig. 9-15. Circuit for Example 9-3, with 200 mA current limiter.

Finally, by switching in different values of R_E or inserting a resistor plus a potentiometer, you can limit the current to different values. This is often desirable when breadboarding experimental circuits.

The complete current limiter for a maximum of 200 mA used in the regulated supply of Example 9-3 is shown in Fig. 9-15.

Although Q2 is usually in saturation, it is possible to have a full supply voltage appear across it if the load should become shorted. When choosing a transistor for Q2, determine the power rating by $P_C = E_{max} \times I_{L\ max}$.

Design Summary

Use the following summary as a guide for designing a current-limiter circuit like the one shown in Fig. 9-16.

1. Determine $R_B = \dfrac{E - 1.4\ V \times \beta}{I_{L\ max}}$

2. Choose $R_E = \dfrac{0.7\ V}{I_{L\ max}}$

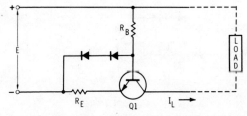

Fig. 9-16. Design parameters for the basic current limiter.

3. Calculate the power dissipated in Q2 if the load shorts, according to

$$P_C = E_{max} \times I_{L\ max}$$

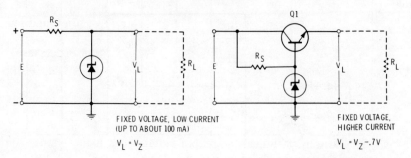

(A) Simple zener regulator.

(B) Zener regulator with series pass transistor.

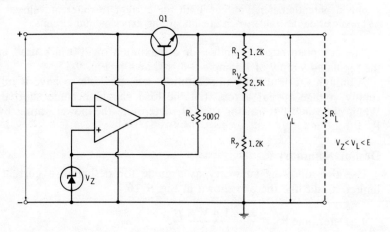

(C) Variable higher current regulator.

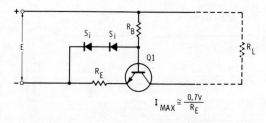

(D) Current limiter.

Fig. 9-17. Summary of regulator circuits.

Quiz

Test your understanding of the variable voltage regulator (Fig. 9-15). Assume $E = 14$ V, V_L is adjusted to 6 V, and $I_L = 100$ mA.

8. If I_L increases, the voltage at point X with respect to ground initially tends to (increase, decrease).
9. This causes the voltage at the slider of R_V to (increase, decrease).
10. This change in voltage at the inverting input of the op amp causes the output of the op amp to (increase, decrease).
11. The resultant (increasing, decreasing) voltage at the base of Q1 finally tends to bring the output at point X back (up, down) to nearly what it was to start with.
12. If you want to adjust the output voltage to a higher value, you would move the slider on R_V (up, down).
13. As drawn, the current limiter limits the maximum current of the supply to 200 mA. If R_E is changed to 7Ω, the maximum output current would be limited to (100 mA, 200 mA, 400 mA).

SUMMARY

Regulators are used to maintain a constant voltage across a load, even if the load current or the input dc voltage changes. A zener diode is the simplest type of voltage regulator. It is used for low-current applications (usually under 100 mA).

Pass transistors can be used as emitter followers to increase the output current capabilities of regulators.

Op amps can be used to provide variable voltage regulators.

Current limiters are useful to protect both the power supply and the test circuit from damage due to overload.

Fig. 9-17 summarizes the various types of regulators discussed in this chapter.

APPENDIX

Quiz Answers

CHAPTER 1

1. (A) $V_p = 1.41\ (V_{rms}) = 1.41\ (12.6) = 18$ V
   (B) If V_r is to be 1 V p-p, the capacitance required is
   $$C \geq \frac{It}{V_r} \geq \frac{(5 \times 10^{-2})\text{ amps} \times (16.7 \times 10^{-3})\text{ sec}}{1.0}$$
   $$\geq 83.5 \times 10^{-5}\text{ F} \geq 835\ \mu\text{F}$$
   (C) Vdcw ≥ 18 V
   (D) 50 mA
   (E) prv $\geq 2\ V_p \geq 36$ V
2. (A) I, (B) D, (C) I, (D) I, (E) D, (F) I
3. Transformer—Solving for the required secondary voltage (rms) and current, we get
   $$V_s = 0.707\ V_o = 0.707 \times 12 = 8.5\text{ V}$$
   and
   $$I_s = 1.81 I_o = 1.8 \times 2 = 3.6\text{ A}$$
   To account for the 1.4 V diode drops, the transformer should have a secondary with about 10 V rms output. The *P5016* seems to fit the requirements.
   Diodes—From Table 1-1, the diodes selected should have a forward current rating of at least $0.5 \times 2.0 = 1.0$ A, and a prv rating of at least 14 V (using the 10 V rms transformer). Diode type *1N4719* can meet these requirements easily.
   Capacitor—The capacitance and working voltage are given by
   $$C = \frac{2.0 \times (8.35 \times 10^{-3})}{2} = 8350\ \mu\text{F}$$

and

$$V_{dcw} = 14 \text{ V}$$

Table 1-4 shows that type *18F2456* has a sufficiently high capacitance and voltage rating. With 8200 µF, the ripple voltage might be too large. You could use two 5500-µF units in parallel, if available, but there would be no savings in cost.

Although the ripple voltage in this problem seems to be a rather high percentage of the output voltage, this power supply would make an excellent input to a regulator circuit (Chapter 9).

4. From Table 1-2, the least expensive transformer having about the right secondary voltage is a 12.6 V filament transformer. The *P8136* will give highly satisfactory results. Output voltage V_o will be approximately $1.41 \times 12.6 \cong 17.8$ V minus the 1.4 V drop across the two diodes, or about 16.4 V.

We need four diodes, each with a forward current rating (I_f) of 0.5 A or more, and a prv of at least 18 V. The *IN4001* would be the most economical choice, although *IN2070* diodes could certainly be used if they are readily available or have a more compact package, etc.

Finally, the minimum filter capacitance required is

$$C = \frac{1.0 \times (8.35 \times 10^{-3})}{3} = 2780 \text{ µF}$$

and from Table 1-4, it appears that the *18F2454* is the best choice. The completed final circuit is shown in Fig. A-1.

CHAPTER 2

1. Defective—you should read a high resistance to both collector and emitter, because both diodes are back-biased.
2. Good—transistor is a pnp type.
3. (A) I—see equation 2-2, (B) I—equation 2-1, (C) D—equation 2-3.

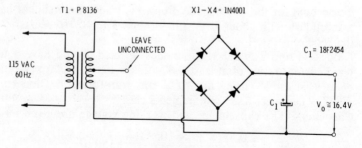

Fig. A-1.

4. (A) S, (B) S—the transistor is a constant current source, (C) I—equation 2-3.
5. (A) S—equation 2-2, (B) D—equation 2-1, (C) I—equation 2-3.
6. (A) S, (B) I—equation 2-3, (C) D—gain varies directly with r_L, (D) D.
7. (A) S, (B) D, (C) I, (D) I.
8. (A) S, (B) S—changing ac load does not affect dc readings, (C) D, (D) D.
9. (A) S, (B) S, (C) I, (D) I.
10. (A) D, (B) D, (C) D, (D) I.
11. (A) I, (B) I, (C) I, (D) I.
12. (A) S, (B) S, (C) D, (D) I.
13. (A) S, (B) S, (C) S, (D) D.
14. All remain about the same.

CHAPTER 3

1. I—the input impedance of Q2 will be higher, making r_{L1} higher and thus increasing the gain of Q1.
2. I—the gain of Q1 will be higher.
3. I—decreasing R_4 causes I_{E1} to be higher, thus increasing the gain of Q1.
4. I—decreasing R_5 turns Q2 on harder.
5. D—this has the opposite effect as decreasing R_5.
6. I—turns Q2 on harder.
7. D—turning Q2 on harder lowers its input impedance.
8. D—decreasing the total ac load on Q1 decreases its gain.
9. $R_E = 1 \text{ V}/0.5 \text{ mA} = 2K$
   $R_C = A \times R_E = 12 \times 2K = 24K$
   $R_2 = 10\, R_E = 20K$
   $R_1 = 17\, R_2 = 340K$

   The voltage across R_1 is about 17 volts, whereas the voltage across R_2 is 1 volt. This is why $R_1 = 17\, R_2$. In the final circuit, you will probably need to make R_1 slightly smaller than this to make up for the 0.3 V base-emitter drop.

10. $R_{E1} \cong \dfrac{R_c}{A_v} \cong \dfrac{5K}{15} \cong 330\, \Omega$

    Then, to keep the emitter current the same as before, the sum of R_{E1} and R_{E2} must still equal 2K. So

    $$R_{E2} = 2K - R_{E1} = 2K - 330 = 1670\, \Omega$$

11. This problem is similar to Example 3-6. The power supply here

calls for only 1.4 Vdc at the emitter, instead of 2 Vdc; otherwise the procedure is the same as in the example, and the result is

$R_1 = 126K, R_2 = 14K, R_3 = 6.3K, R_4 = 100 \Omega,$
$R_5 = 1.3K, R_6 = 126K, R_7 = 14K, R_8 = 6.3K,$
$R_9 = 1.4K, R_{10} = 10K.$

Remember that these values are calculated as a starting point—the circuit will work just as well with the closest standard values. Resistors R_4 and R_{10} have the greatest effect on circuit gain.

12. Better—see equation 3-6.
13. The same—size of coupling capacitors does not affect f_2.
14. Worse—equation 3-6 (R_t is reduced).
15. Better—equation 3-7 (R_{eq} is reduced).
16. The same—transistor capacitance does not affect f_1.
17. Better—equation 3-7.
18. The same—small shunt capacitance does not affect f_1.
19. Worse—the total shunt capacitance is greater, causing high frequencies to be shunted to ground. This is an effective way of "rolling off" the high end to reduce noise or avoid some undesired frequency. By trying different values of C_s, you can get the high frequencies to roll off almost anywhere you want.
20. Fig. 3-11—this circuit will have a high input impedance and a stable gain. Since space is of some concern, you would probably use only one stage.
21. Fig. 3-4—this circuit has the highest gain for a compact amplifier. More than one stage would probably be needed for a gain of 2000.
22. Fig. 3-14—Here you would use the two-stage circuit with feedback to get stability. A single stage with feedback would probably not give enough gain.
23. Either Fig. 3-5 or Fig. 3-9—the Darlington circuit will give an even higher input impedance than the single-stage emitter follower, so it might be a better choice.
24. Fig. 3-3—this circuit has the highest gain for a single stage. As with any circuit, there is no single *best* design for these applications, but merely a few good compromises based on several factors. The examples given here are guidelines to help you make an acceptable choice.

CHAPTER 4

1. Correct—the transistor is a p-channel JFET.
2. 2 MΩ—since virtually zero current flows into the gate, $r_{in} \cong r_g$.
3. Decrease—this will reduce the bias, cause more drain current to flow, and cause a greater drop across R_D.

4. 300 mV—the solution is arrived at according to
$$A_v \cong y_{fs} \times R_D \cong (3 \times 10^{-3}) \times (5 \times 10^3) = 15$$
and
$$V_o = A_v \times V_{in} = 15 \times 20 \text{ mV} = 300 \text{ mV}$$

CHAPTER 5

1. Calculation of V_o is done by the method given in Example 5-2. In this cause, however, the equations give $V_o = 40$ V, which is much greater than the normal power supply voltage (± 15 V). The amplifier is in saturation, so the actual value of V_o measured will be approximately 15 V.
2. The solution of this problem requires that we use the general expression for V_o (equation 5-5). If $R = R_1 = R_2 = R_3$, then equation 5-5 simplifies to

$$V_o = -(V_1 + V_2 + V_3)\frac{R_F}{R}$$

$$V_o = -9 \text{ V} \frac{5K}{10K} = -4.5 \text{ V}$$

By adjusting the various values of R and the value of R_F, you can use this circuit to add or average any number of inputs. Accuracy will generally be good as long as V_o does not approach the plus or minus supply voltage.

3. 3.0 V/sec—this solution is obtained by the method described in Example 5-7.
4. 900 mV—the method is outlined in Example 5-8. Note that, if $R_{F1}/R_1 \neq R_{F2}/R_2$, the voltage gain for each input will not be the same, and there will be some output even when $V_1 = V_2$. Another way to view this is to say that the cmrr of the amplifier will be maximum only when R_{F1}/R_1 and R_{F2}/R_2 are exactly equal. An easy way to optimize the circuit is to make one of the four resistors variable with a pot. Apply about 3 Vdc to both inputs—points A and B—simultaneously (using a large series resistance of, say, 100K to reduce current flow), then adjust this gain pot until the dc output of the op amp is nulled to zero. The circuit is now "balanced," and will reject signals applied to both inputs simultaneously, such as 60-Hz noise.

CHAPTER 6

1. $V_{oo} = \dfrac{R_1 + R_F}{R_1} \times V_{io} = \dfrac{2K + 10K}{2K} \times 4 \text{ mV} = 24 \text{ mV}$

2. $V_{oo} = \dfrac{40K + 200K}{40K} \times 4\text{ mV} = 24\text{ mV}$

3. $V_{oo} = \dfrac{10K + 200K}{10K} \times 4\text{ mV} = 84\text{ mV}$

   Notice that the output offset voltage is the same in problems 1 and 2 because, although the resistances are different, the amplifier *gain* is identical in both cases.

4. $R_1 = Z_{in} = 2K$

   $R_F = A_v \times R_1 = 25 \times 2K = 50K$ (equation 5-8)

   $R_2 = \dfrac{R_1 R_F}{R_1 + R_F} = \dfrac{2K \times 25K}{2K + 25K} = 1.85K \cong 1.8K$

5. True—R_2 is equal to the parallel combination of R_1 and R_F.
6. False—If the circuit gain is not very high so that V_{io} is negligible, and if R_2 is used to compensate for bias currents, the output offset will be quite small.
7. $f_{max} = 1$ kHz—see Fig. 6-13.
8. $f_{max} \cong 300$ kHz—see Fig. 6-11.
9. $R_1 = 0$, $C_1 = 10$ pF, $C_2 = 3$ pF—Fig. 6-11.
10. 741—no compensating components required.
11. The highest frequency to be amplified is 400 Hz with a gain of 100, and Fig. 6-13 shows that this is well within the range of a 741. Next, we must have a constant input impedance of 500 ohms for the resistor connected to the inverting input. Since the gain must be variable, we can use a variable feedback resistor. The minimum value of R_F should be 500 ohms for a gain of unity, and a maximum value of 50K, for a gain of 100. A 50K pot in series with a fixed resistor of 500 ohms will do the job. The final circuit is shown in Fig. A-2. No null control is necessary unless the output must be exactly zero when the input is zero.
12. For a high-frequency response and high gain, the externally-compensated 709 device should be used. Next, we need an input impedance of at least 100K, which calls for a noninverting amp. It would be possible to use a 100K resistor to the input of an inverting amp, but that would mean that R_F would have to be about $1000 \times 100K = 100$ MΩ. With such a large value of R_F, there would probably be excessive output due to the offset current.

    From Fig. 6-11 a gain of 1000 (60 dB) requires that the compensation components be $R_1 = 0$, $C_1 = 10$ pF, and $C_2 = 3$ pF. The final circuit is shown in Fig. A-3. Different values for R_1 and R_F could be used, as long as the ratio of R_F/R_1 is the same. Remember, too, that using very large values of R_F (several megohms) will cause more problems with input offset current.

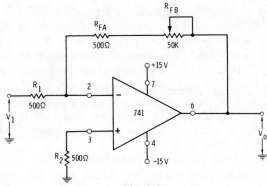

Fig. A-2.

CHAPTER 7

1. $R = \dfrac{1}{18\,Cf} = \dfrac{.055}{555\,(10^{-8})} = 10^4 = 10K$

2. $R = \dfrac{1}{2\pi Cf} = \dfrac{.16}{(10^{-8})(1600)} = 10K$

3. (A) $V_{CC} = 9$ Vdc.
   (B) Choose $R_1 = R_2 < R_x < 500K$. (Any value less than this will work—let $R_1 = R_2 = 100K$.)
   (C) Solving for C,
   $$C = \dfrac{1}{1.4\,R_1 f} = \dfrac{.715}{R_1 f} = \dfrac{.715}{(10^5)(715)} = 10^{-8} = 0.01\ \mu F$$

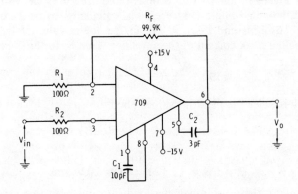

Fig. A-3.

4. Assuming we take the output from the collector of Q2, as shown in Fig. 7-8, the requirement for a 10% duty cycle is that $R_2C_2 = 0.1\ (R_1C_1)$. In problem 3, we used $R_1 = R_2 = 100K$, which means C_2 would need to be changed to one-tenth of C_1, or $C_1 = 0.01\ \mu F$, $C_2 = 0.001\ \mu F$. The same result is obtained if we leave $C_1 = C_2 = 0.01\ \mu F$ and make R_2 equal to 1/10 of R_1, or $R_1 = 100K$, $R_2 = 10K$. Either method will give a 10% duty cycle measured at the collector of Q2. Note that, since Q1 is switched on whenever Q2 is switched off, the duty cycle measured at the collector of Q1 will be 90%.
5. Essentially all the oscillator current passes through collector resistors R_x and R_y (because R_1 and R_2 are larger by approximately a factor of β). Also, R_x is conducting current whenever R_y is not—at any given time, current is flowing through either R_x or R_y, but not through both; so if we let $R = R_x = R_y$, the current required by the circuit is

$$I \cong \frac{V_{CC}}{R} \cong \frac{9\ V}{10K} \cong 1\ mA$$

CHAPTER 8

1. As shown in Example 8-1,

$$I = 560\ mA = \sqrt{\frac{P}{R_L}}$$

Next, solving for P and substituting $R_L = 8$ ohms, we get

$$P = I^2 R_L = (.560)^2\ (8) = .312\ (8) \cong 2.5\ W\ rms$$

Note that, if I_C is held constant, the power consumed is proportional to R_L.

2. (A) The solution to this problem requires that we work backwards from Example 8-1. V_{max} was calculated to be 12.6 V with $R_L = 16$ ohms and $I_{C\ max} = 790$ mA; a lower value of R_L will increase the peak current of the amplifier. That is, for an 8-ohm load,

$$I_{C\ max} = \frac{V_{max}}{8} = \frac{12.6}{8} \cong 1.58\ A$$

which is twice the previous value. Then

$$I_{rms} = I_{C\ max}\ (.707) = 1.12\ A$$

and

$$P = I^2 R = (1.12)^2\ (8) = 10\ W\ rms$$

Clearly, then, using half the load impedance in this case results in a doubling of I_C and a doubling of output power.

(B) Probably it does—even if the circuit is not damaged, it will be operating under an extreme overload. At least some redesigning is necessary to accommodate the extra heat dissipated, etc. The design areas most affected are the power transistors, load resistors, and the choice of heat sinks (see next section of Chapter 8).

(C) There are two easy options here that involve no modification of the amplifier circuit. The first is to reduce V_{CC} to about 15-20 Vdc, which will lower I_C and reduce the power to approximately what it was before (5 W). Alternatively, you can simply wire two 8-ohm speakers in series across the output; the speakers in series effectively produce a 16-ohm load impedance.

3. $\theta_{SA} \leq 7.5°C/W$. The method used is the same as for Example 8-2.
4. $\theta_{SA} \leq 8.0°C/W$, using equation 8-2 where θ_{CS} is omitted.
5. No—for most equipment designed to be used indoors, 50°C will probably be a safe maximum value. However, the designer should always visualize the intended purpose of the circuit and try to foresee any problems. As one example, the interior of an automobile occasionally exceeds 150°F (65°C) on a hot summer day. Obviously, an amplifier intended for outdoor or mobile use should be able to withstand such temperatures. Indoors, be sure the heat sink is kept well away from appliances that may also be generating considerable amounts of heat. There are many other instances you can think of; keep them in mind when selecting and installing components for the circuits you design.

CHAPTER 9

1. Voltage.
2. Low—zeners are not very efficient for high-current applications.
3. Come out of conduction when E drops to its minimum value.
4. E is maximum—I_s will be maximum.
   I_L is minimum—I_z will be maximum.
5. E is minimum.
   I_L is maximum—R_s must be small enough to allow some current to flow through the zener.
6. E is maximum.
   I_L is maximum.
7. Slightly lower—there will be approximately 0.7V drop across the base-emitter junction.
8. Decrease—in general, any increase in load current drawn from a supply tends to decrease the supply voltage.
9. Decrease—Resistors R_1, R_v, and R_L form a voltage divider across the load.

10. Increase.
11. Increasing, up—The pass transistor acts as an emitter follower.
12. Down—again, making the inverting input less positive makes the output more positive.
13. 100 mA—with 100 mA flowing through 7Ω, the drop across the resistor will be 0.7V. The voltage across R_E cannot rise above 0.7V because of the diodes connected between the base and emitter of Q_2.

Index

Index

A

Ac resistance, 51
Action, diode, 9
Amplification, signal, 36-37
Amplifiers
  audio power, 143-155
    biasing, 144-145
    design summary for, 151-152
  common-emitter, 36-42
  complementary-symmetry, 143-152
    biasing, 144-145
    design summary for, 151-152
  differential, 93-99, 109-111
  frequency response of, 64-67
  invert, 103-104
  JFET, 78-83
  noninvert, 104-105
  operational, 100-111
Applications, FET, 87-90
Attenuator, voltage-controlled, 90

B

Beta (β), 32-34
  definition of, 32
Bias currents, op-amp, 109
Bias current, input, 109, 115-117
Bias, transistor, 31-36
Bias, voltage divider, 39-43, 46
Biasing audio amplifiers, 144-145
Biasing JFETs, 78-81
Bipolar transistors, 29-31
Breakdown, zener, 157-158
Bridge, full-wave, 19-22
Bypass capacitors, emitter, 65-66

C

Calibration, voltmeter, 88-89
Capacitive reactance, 65-66
Capacitance, input, 66-67
Capacitor
  coupling, 36, 56
  emitter bypass, 65-66
  filter, 11-14
  grounding, 41
  working voltage of, 13-14
Channel, FET, 76-77
Chopper, FET, 90
Circuits
  Darlington, 57
  gain of, 50-58
  integrated, 101-102
  regulator, summary, 172-173

Closed loop gain, 59-61
Common-emitter amplifier, 36-42
  design summary for, 42-43, 45-46
Common-mode rejection, 99
  ratio, 99
Common-mode signal, 99
Complementary-symmetry amplifiers, 143-152
  biasing, 144-145
  design summary for, 151-152
Constant-current source, 35, 93-96
  JFET, 79
Control
  gain, 68-70
  offset null, 114-116
  volume, 68-70
Converter
  sine-wave to square-wave, 137-138
  square-wave to triangular-wave, 137-138
Coupling capacitor, 36, 56, 71
Crossover distortion, 144-145
Current
  input bias, 115-117
  input offset, 117
  leakage, 93-94
  limiters, 169-173
  op-amp, 109
  source, constant, 35
Cutoff
  frequency, 67
  transistor, 35, 37

D

Darlington circuit, 57
Depletion, 83
  -enhancement, 84
  region, 76-77
Design summary
  audio amplifier, 151-152
  power supply, 22-24

Differential amplifiers, 93-99, 109-111
Differential signal, 109-111
Diode(s)
  action, 9
  identification numbers, 8
  junction, 8
  prv of, 14
  regulator, zener, 157-160
  solid-state, 7-10
DIP package, 120-123
Distortion, crossover, 144-145
Drain, 76

E

Effects, temperature, 44-45
Emitter bypass capacitors, 65-66
Emitter follower, 53-57, 88-89
  in tvm, 89
Enhancement, 84
External frequency compensation, 118-119

F

Feedback in amplifiers, 58-63
FETs, 75-92
  applications, 87-90
  channel, 76-77
  chopper, 90
  source follower, 87
  terms, glossary of, 92
Filament transformers, 15
Filter capacitor, 11-14
Follower
  emitter, 53-57
  source, FET, 87
  voltage, 106
Frequency
  compensation, 118-120
    external, 118-119
    internal, 119-120
  cutoff, 67

Frequency—cont
 response, 117-120
  amplifiers, 64-67
 Full-wave bridge, 19-22
 Full-wave rectifier, 18-19

G

Gain, 59-61, 70
 circuit, 50-58
 closed loop, 59-61
 controls, 68-70
 FET, 81-82, 90
 op-amp, 104, 105, 109
 open loop, 61
 transistor, 37-38
Gate, 76
Generator
 linear ramp, 107-109
 sine-wave, 127-132
 square-wave, 132-136
 waveform, 127-141
  summary, 139-141
Germanium transistor, 40
Glossary
 FET terms, 92
 op amp terms, 124-125
Grounding capacitor, 41

H

Half-wave rectifier, 10-17
Heat sinks, 152-154
h_{FE} (see Beta)
Higher current regulators, 161-165

I

Identification numbers
 diode, 8
 transistor, 29
Impedance, 54-60
 input, 47-53
 measuring, 47-52

Input
 bias current, 109, 115-117
 capacitance, 66-67
 impedance, 47-53, 85
  JFET, 75, 80-83
  measuring, 47-52
 offset current, 117
 offset voltage, 113-115
 resistance, FET, 90
Integrated circuits, 101-102
Integrator, 107-109
Internal frequency compensation, 119-120
Inverting amplifier, 103-104

J

JFET, 75-92
 amplifier, 78-83
 biasing, 78-81
Junction
 diode, 8
 transistor, 29-31

L

Leakage current, 93-94
Limiters, current, 169-173
Linear ramp generator, 107-109

M

Measuring input impedance, 47
Midband, 65
Model, sandwich, 29-30
MOSFET, 75-92

N

Noninverting amplifier, 104-105
Null control, offset, 114-116

Numbers
  diode identification, 8
  transistor identification, 29

O

Offset
  current, input, 117
  null control, 114-116
  voltage
    input, 113-115
    output, 113-117
Ohmmeter test, transistors, 30-31
Op-amp
  bias currents, 109
  gain, 104, 105, 107
  terms, glossary of, 124-125
Open loop gain, 61
Operational amplifiers, 100-111
Oscillator(s), 127-141
  phase-shift, 127-130
    twin-T, 130-132
    square-wave, 132-136
Output offset voltage, 113-117

P

Package
  DIP, 120-123
  TO-99, 120-123
Phase-shift oscillator, 127-130
Power amplifiers
  audio, 143-155
  biasing, 144-145
  design summary, 151-152
Power supply
  design summary, 22-24
  regulated, 157-173
  transformers, 14-17

R

Ramp generator, linear, 107-109

Ratio, common-mode rejection, 99
Reactance, capacitive, 65-66
Rectifier
  half-wave, 10-17
    full-wave, 18-19
Region, depletion, 76-77
Regulated power supplies, 157-173
Regulator(s)
  circuits, summary, 172-173
  higher current, 161-165
  series, 162-165
  variable voltage, 165-169
  zener diode, 157-160
Rejection, common-mode, 99
Resistance ac, 51
Response
  frequency, 117-120
    amplifiers, 64-67
Ripple voltage, 11-13
Rms voltage, 15

S

Sandwich model, 29-30
Saturation, 37
  JFET, 80-81
  transistor, 35
Series regulator, 162-165
Signal
  amplification, 36-39
  common-mode, 99
  differential, 109-111
Sine-wave
  generator, 127-132
  -to square-wave converter, 136-137
Sinks, heat, 152-154
Solid-state diodes, 7-10
Source, 75-76
  constant current, 35, 93-96
  follower, FET, 87
  -to triangular-wave converter, 137-138

Square-wave
  generator, 132-136
    with nonsymmetrical output, 134-136
Summary
  design
    audio amplifier, 151-152
    regulator circuits, 172-173
    waveform, generators, 139-141
  Supplies, power, regulated, 157-173

T

Temperature effects, 44-45
Terms
  FET, glossary of, 92
  op amp, glossary of, 124-125
TO-99 package, 120-123
Transformers
  power-supply, 14-17
  filament, 15
Transistor(s)
  beta (β), 32-34
  bias, 31-36
  bipolar, 29-31
  cutoff, 35
  gain, 37-38
  germanium, 40
  identification numbers, 29
  junction, 29-31
  ohmmeter test of, 30-31

Transistor(s)—cont
  saturation, 35
Twin-T oscillator, 130-132

V

Variable voltage regulators, 165-169
Voltage
  -controlled attenuator, 90
  divider bias, 39-43, 46
  follower, 106
  input offset, 113-115
  output offset, 113-117
  regulators, variable, 165-169
  ripple, 11-13
  rms, 15
  working, capacitor, 13-14
Voltmeter
  calibration, 88-89
Volume controls, 68-70

W

Waveform generators, 127-141
Working voltage, capacitor, 13-14

Z

Zener breakdown, 157-158
Zener diode regulator, 157-160